Seeding Empire

Seeding Empire

AMERICAN PHILANTHROCAPITAL
AND THE ROOTS OF THE GREEN
REVOLUTION IN AFRICA

Aaron Eddens

UNIVERSITY OF CALIFORNIA PRESS

University of California Press
Oakland, California

© 2024 by Aaron Eddens

Library of Congress Cataloging-in-Publication Data

Names: Eddens, Aaron, 1985- author.
Title: Seeding empire : American philanthrocapital and the roots of the
   green revolution in Africa / Aaron Eddens.
Description: Oakland, California : University of California Press,
   [2024] | Includes bibliographical references and index.
Identifiers: LCCN 2023048603 (print) | LCCN 2023048604 (ebook) |
   ISBN 9780520395299 (cloth) | ISBN 9780520395305 (paperback) |
   ISBN 9780520395329 (epub)
Subjects: LCSH: Green Revolution—Africa. | Economic assistance,
   American—Africa. | Transgenic plants—Africa. | Climatic changes—
   Africa. | Green Revolution—History.
Classification: LCC S472.A1 E43 2024 (print) | LCC S472.A1 (ebook) |
   DDC 338.1096—dc23/eng/20231212
LC record available at https://lccn.loc.gov/2023048603
LC ebook record available at https://lccn.loc.gov/2023048604

33 32 31 30 29 28 27 26 25 24
10 9 8 7 6 5 4 3 2 1

For Dad, in loving memory

Contents

	List of Figures	ix
	Acknowledgments	xi
	Introduction: Biotech Agriculture's Final Frontier	1
1.	How We Remember the Green Revolution	16
2.	"A Green Revolution, This Time for Africa"	38
3.	"The Landraces Are in the Hybrids"	62
4.	Seeing Like a Seed Company	86
5.	Securitizing Smallholder Farmers on the Front Lines of the Climate Crisis	107
	Conclusion: What Can the Green Revolution Teach Us about Climate Change?	127
	Notes	141
	Bibliography	169
	Index	185

Figures

1. Norman Borlaug: "The Man Who Saved a Billion Lives" banner, Des Moines, Iowa — 17
2. Norman Borlaug with trainees, Mexico — 26
3. Alliance for a Green Revolution in Africa, West End Towers, Nairobi, Kenya — 42
4. Gates Foundation Discovery Center, Seattle, Washington — 55
5. The Rockefeller Foundation Agricultural Survey Commission, Mexico, 1941 — 66
6. Students at the School of Huichapan, Mexico, 1941 — 69
7. Examination of indigenous varieties of corn, Mexico, 1959 — 81

Acknowledgments

This book would not be in your hands today without the support, guidance, and love I have received from the following people. I'm honored to be able to express my gratitude. If anything, I hope this book provides an opportunity for more conversations with those I thank below—and sparks new ones with many others.

To begin, I want to thank two outstanding mentors I had as an undergraduate at Western State College in Gunnison, Colorado. John Hausdoerffer's environmental studies classes first piqued my interest in thinking about nature and power. And Christy Jespersen's environmental justice literature class opened lines of questions that I'm still following.

This book began as a dissertation in the Department of American Studies at the University of Minnesota. Rachel Schurman's thoughtfulness and enthusiasm for learning with others motivated me from our first conversation. She taught me how to chase down interviews and never hesitated to tell me when my writing was as "clear as mud." Bianet Castellanos pushed me to think big, while also showing me the importance of life outside the academy. The form this book eventually took owes much to her guidance. Tracey Deutsch encouraged me to think like a historian. And Susan Craddock pushed my thinking at just the right time. I also learned much from seminars and conversations with Jennifer Pierce, Kevin Murphy, Erika Lee, Saje Mathieu, Abby Neely, Arun Saldanha, Will Jones, Juliana Hu Pegues, Colin Agur, Catherine Squires, and George Henderson.

Thanks to argi-food reading group friends: Ilona Moore, Valentine Cadieux, Rachel Slocum, and Stephen Carpenter. A fellowship at the University of Minnesota's Institute for Advanced Study planted seeds that bore fruit in this book. Thank you, Jennifer Gunn, Brianna Menning, Karen Kinoshita, Christina Collins, Susannah Smith, Sami Poindexter, Madison Van Oort, David Lemke, Jen Hughes, Amber Annis, and Sarah Saddler. I want to thank friends and colleagues that provided much-needed laughs and support at Minnesota and beyond: Heidi Zimmerman, Ben Wiggins, Matthew Schneider-Mayerson, the dearly missed, Jesús Estrada-Pérez, Joe Whitson, Matt Boynton, Sarah Atwood-Hoffman, Rose Miron, Mingwei Huang, Katy Mohrman, Joe Getzoff, Kai Bosworth, Emily Springer, Bernadette Pérez, Erik Kojola, Laura Matson, Kirsten Anderson, Jessie Anderson, Alex Liebman, Rachel Vaughn, Michelle Yates, Shane Hall, Matt Huber, Erica Hannickel, Jay Fiskio, Alex Chambers, Shannon Mancus, Sarah Wald, Becca Ballard, Louise Davis, Bob Johnson, Garrett Graddy-Lovelace, and Hossein Ayazi.

Since joining the faculty at Grand Valley State University, I have had the privilege to work with creative, supportive colleagues. Thanks to my colleagues in the Brooks College of Interdisciplinary Studies. In addition, a special thanks to Jennifer Cathey, Mary Wiliford, Jack Mangala, Coeli Fitzpatrick, Gamal Gasim, Denise Goerisch, Jen Jameslyn, Pat Johnson, Jakia Marie, Julia Mason, Andrew Schlewitz, Melanie Shell-Weiss, Joel Stillerman, Kim McKee, Mark Schaub, Bren Tooley, and Meredith Fedewa. I also want to thank the terrific students with whom I get to learn alongside every day.

This project benefitted from several sources of funding. I would like to thank the Rockefeller Archive Center, the American Studies Association, and the Agricultural History Society for research and conference travel funding. At the University of Minnesota, I received support from the Mark and Judy Yudoff Fellowship, the Interdisciplinary Center for the Study of Global Change, the Department of American Studies, the Council of Graduate Students, the Institute for Advanced Study, and the Graduate School. At GVSU, I have received generous support from the Center for Scholarly and Creative Excellence, including a course-release grant in winter 2023 that helped provide much-needed writing time as I completed the book.

Parts of chapter 3 appeared previously as an article in the *Journal of Peasant Studies* and parts of chapter 5 were published in an article in *American Studies*. I thank both journals for permission to reproduce portions of those articles.

Several people helped me move the book toward completion. Evan Taparata read working drafts and his numerous suggestions helped me make crucial revisions. At GVSU, Max Counter, Daniela Marini, and Ramya Swayamprakash talked with me about the writing process and helped to boost morale as I was finishing. Badia Ahad's book proposal workshop was incredibly helpful. Niels Hooper enthusiastically supported the project and Nora Becker has cheerfully guided me through the submission process. I deeply appreciate the insights from David Roediger and an anonymous peer reviewer, which helped me to clarify my goals and sharpen my arguments. Emily Park and Jon Dertien patiently guided me through the production process. And Gary Hamel's careful copyediting was much appreciated.

I also owe a debt to people who helped me during research trips. I would like to thank Carey and Susan Curelop, Catherine Kuzmicki, and Geoffrey Njuguna. I would also like to thank all of my interviewees for offering their time and insights. Erin George at the University of Minnesota Archives helped me navigate the Green Revolution records, and Lee Hiltzik, Mary Ann Quinn, and Renee Pappous at the Rockefeller Archive Center were terrific.

My mom and dad deserve more thanks than I can put into words for inspiring me to be a reader and writer. Thanks, Mom, for teaching me how to write and for always encouraging me to use my voice. And Dad: we miss you every day. But I often think of your love for books and bookstores and know that you instilled in me a strong desire to think critically. Thanks also to my grandma, Beverley Curelop, for always asking me what books I was reading. I want to acknowledge the long-distance encouragement of my Texas family: Jason Eddens and Morgan Bathe, Callie and Vance Tillman, and Kyle Eddens. My Wisconsin family, Tom and Marilyn Nyre, have steadfastly supported us. Thanks also to Erik Nyre and Catherine Kuzmicki. I hope we can celebrate together soon.

Finishing this book has reminded me what matters most. So I save my last thank-yous for Emily Nyre and Rita and Marlowe Eddens. Emily: I am so grateful for your love, your humor, and your strength. You inspire me every day and I can't wait to write more chapters with you. Rita, your creativity and boundless joy have motivated me throughout this book's long journey. I'm excited to read the books you'll soon write. And Marlowe, your happiness and sense of wonder has kept me grounded. Thanks for not letting any of us take ourselves too seriously.

Introduction

BIOTECH AGRICULTURE'S "FINAL FRONTIER"

In October of 2009, Bill Gates, the Microsoft founder and co-chair of the Bill and Melinda Gates Foundation, gave the keynote address at the World Food Prize conference in Des Moines, Iowa. At the time, the event—an annual gathering of hundreds of the most influential people in international agricultural development—seemed an unlikely venue for a speech from the world's most famous techie and billionaire philanthropist.[1] As Gates took the podium in front of the jam-packed ballroom of Des Moines's downtown Marriott, he told the audience that he and Melinda had recently become passionate about improving the lives of poor farmers in the Global South. Because they were new to the subject, he explained, they had been inviting various experts to their foundation to teach them about global agriculture. In all these conversations, he recounted, they kept hearing about one person: Norman Borlaug.

Gates outlined how Borlaug became central to American-led development projects across Asia and Latin America known as the Green Revolution. While working for a Rockefeller Foundation agricultural program in Mexico in the 1960s, Borlaug developed a high-yielding variety of wheat that catalyzed record-setting yields across India and Pakistan. In the process, he refuted Neo-Malthusian doomsayers that had warned of famines across Asia. He was awarded a Nobel Peace Prize in 1970. Summarizing the scientist's legacy, Gates said that Borlaug's Green Revolution had "helped divert famine, save hundreds of millions of lives, and lift whole countries out of poverty." Gates's audience knew this story well. Indeed, Borlaug had founded the World Food Prize and his legacy is frequently celebrated at the event.[2] Had he not died just before the 2009 conference, he would have been eagerly listening to Gates from his usual front-row seat. Yet as Gates memorialized the conference's central figure, he stressed that Borlaug's work was unfinished. Though Gates called the Green Revolution "one of the great achievements of the twentieth century," he argued that it had failed on one crucial front: "it didn't go to Africa."

Describing Africa as woefully behind other continents in terms of per capita crop yields, Gates declared that the time was ripe for a Green Revolution on the continent—an effort his foundation would support through a suite of new grants. This Revolution, he argued, would focus on the needs of the world's "smallholder farmers," the millions of farmers that toil on small plots of land and are largely disconnected from international commodity markets. With climate change making the plight of smallholders increasingly vulnerable, Gates insisted that Western agricultural technologies like genetically modified (GM) crops would play a pivotal role in the new Green Revolution—and announced that his foundation was already attempting to develop drought-tolerant, GM crops for Africa's smallholders.

But, Gates warned, this mission faced a tremendous challenge: overcoming the opposition of Westerners that opposed GM crops because of health or environmental concerns. The normally mild-mannered Gates denounced biotech opponents with uncharacteristic zeal:

> They've tried to restrict the potential use of biotechnology in Sub-Saharan Africa without regard to how much hunger and poverty might be reduced by it or what the farmers themselves might want. Some voices are instantly hostile to any emphasis on productivity. They act as if there's no emergency—even though, in the poorest, hungriest places on Earth, population is growing faster than productivity and the climate is changing.[3]

This was also a familiar story among the Des Moines crowd. Officials from the world's largest agricultural biotechnology companies, DuPont Pioneer and Monsanto, nodded in agreement as Gates reiterated an argument they had voiced for years. Since introducing commercial genetically modified organisms (GMOs) in the 1990s, they had argued that poor farmers in the Global South could benefit from their GM seeds—if only anti-GMO activists in the North would get out of the way. Africa had been the locus of their concern. At the time of Gates's speech, only four African countries permitted any GM agriculture. Blaming "privileged" Westerners, biotech advocates argued the continent was "starved for science."[4] In the last decade of his life, Borlaug himself penned a series of editorials condemning "anti-science zealots" for keeping GMOs out of the hands of poor farmers in Africa. Hearing the world's richest man—and co-chair of the most influential global philanthropy—echo Borlaug filled the room with a palpable excitement. As Gates concluded his remarks, the audience erupted into a "spontaneous standing ovation" the likes of which had never been seen in the conference's twenty-five-year history.[5]

Since Gates's speech in Des Moines, The Gates Foundation, the US Agency for International Development (USAID), African governments, multilateral public sector research institutions, and multinational agribusiness corporations have joined forces to transform farming systems across Africa. Gates's call to extend Borlaug's Green Revolution with biotech crops would drive a growing number of development partnerships aiming to develop new GM varieties and change regulatory policies across the continent. This book, though, is not primarily about the battle to extend biotech crops to what has been called their "final frontier." Instead, it offers a "genealogy" of the ideas underpinning a longer history of Green Revolution projects—from its roots in Borlaug's program in Mexico in the 1940s and 1950s through its Cold War–era international expansion to today's burgeoning Green Revolution in Africa.[6] I show how the projects across this lineage share common narratives that center the agency of White Westerners who bring technology to the agrarian frontier while devaluing the knowledge of indigenous people and smallholder farmers the world over. These logics shore up geographical and historical ideas that draw sharp lines between those deemed most vulnerable and those tasked with saving them. In so doing, they distance the source of the problem to countries in the Global South—and refuse to think relationally about the root causes of global inequalities.

The story about the Green Revolution's future is also about its past. Gates's use of Borlaug makes this clear. But Gates's speech also demonstrates that history is not just the past. It is the stories we tell about the past. Gates describes Borlaug's Green Revolution as a straightforward history of American scientific innovation overcoming hunger and poverty in the Global South. But this history is decidedly more complicated than Gates's neat narrative suggests. The Green Revolution might have increased food supply, but hunger and poverty persist in the countries in which it was most active, largely because of economic inequalities. Gates also glosses over more of the historical complexities and contradictions of the Borlaug success story. A short list of these would include the Green Revolution's ties to pesticide poisoning in indigenous communities, the exacerbation of rural inequalities, and its role in displacing farmers from their land and toward urban slums. And yet Gates's version of the Green Revolution is the one usually told across university, government, industry, and NGO circles.[7] Despite substantial critiques of the Green Revolution from scholars and civil society groups, its power as a narrative persists. Indeed, in the days following Gates's speech, some of the world's most powerful people repeated it in their calls to transform agriculture across Africa.[8] I argue that to understand the Green Revolution's remarkable persistence, we need to broaden our questions about its roots: How are African geographies conceptualized in the Global North as sites of "emergency"? How might contemporary efforts to introduce Western technologies to smallholder farmers in Africa reproduce long-standing racial logics that dehumanize non-Western Others as "not-yet" developed? And how do contemporary discourses around food security and climate change rewrite older scripts about poverty, hunger, and security?

Seeding Empire pursues these questions. As the title suggests, I foreground the analytic of empire. Historians have shown how the Green Revolution in Borlaug's era was fundamentally tied to US foreign policy. The United States saw peasant farmers across Asia and Latin America as the front lines in a global battle between freedom and communism. The nongovernmental institutions in today's Green Revolution might not be as deeply entwined with US state power as the Rockefeller and Ford Foundations were during the Cold War. But that does not mean that the United States has taken a backseat. From massive USAID investments to catalyze the private maize seed sector to the State Department's role in funding biotechnology promotion to the increasing alignment of the "development, diplomacy, and defense" pillars of American national security strategy, US power remains crucial to defining the parameters of the Green Revolution in Africa.[9] Inspired by scholarship that examines the "coloniality" of American agriculture, this book makes connections across an American empire that is "always in process" and its projection of "soft power" around the globe.[10]

To insist on thinking with empire means looking not only at the entanglements of state and capital in frontier-making development projects, but also at the project of empire in relation to history.[11] The Green Revolution offers an example of a dominant narrative about US history that maintains its power by obscuring its ties to American empire. Revealing these connections yields a better understanding of the enduring power of stories about Borlaug and the Revolution's past. It also demonstrates how that history has been used to drive contemporary agricultural development in Africa. Suggesting that the Green Revolution in Africa is part of US empire, however, runs the risk of reproducing a kind of imperial logic. Overemphasizing the role of the United States can simplify the multidirectional forms of power across different locations and scales of the Green Revolution. I do not intend to reduce the Green Revolution in Africa to a singular story of a kind of capital E Empire. Nonetheless, as I show throughout this book, empire offers an effective analytical framework for tracing the intersecting modalities of power forged through the ongoing Green Revolution. In the context of debates about how to develop new ways of thinking internationally to address the climate crisis, we must grapple with the persistence of empire.

THIS BOOK'S APPROACH TO STUDYING THE GREEN REVOLUTION

Two years after Gates's speech in Des Moines, I had my own experience with the Green Revolution. It was during my first semester of graduate studies at the University of Minnesota. I had joined a weekly seminar of faculty and graduate students engaging with the politics of global agriculture. As it happened, the University Archives had recently digitized a large collection of documents, including the papers of its most famous agricultural alumnus: Norman Borlaug. They were celebrating the project with an exhibit on the "Minnesota Roots of the Green Revolution." On a breezy September afternoon, our seminar group went to check it out.

I remember walking through the collection of Borlaug's material: photos, newspaper clippings, and leather-bound field notebooks with the scientist's handwritten notes about his wheat breeding efforts in Mexico. Aside from a few newspaper articles about his support of the notorious agri-chemical DDT, the exhibit portrayed Borlaug as a dedicated scientist that overcame difficult working conditions to achieve transformative results. This celebratory account was consistent with the larger narrative the University told in its homages to Borlaug, especially on the St. Paul campus, where one can walk into Borlaug Hall and find a few "Ag" folks who remember working with "Norm." Yet our group of historians and social scientists had been having a much different conversation about the Green Revolution, one that complicated the untarnished account of Borlaug the miracle-worker. We had been reading about the intersections of US power, scientific hubris, and agribusiness expansion that this hero narrative obscures. Despite this academic evidence that the Borlaug story was, at the very least, more complicated than its usual telling, the version that Gates recounted in Des Moines remained unscathed. Around this same time, Gates and others were rehearsing the Borlaug tale in their calls to transform agriculture across Africa. When President Barack Obama announced a new global partnership aimed at jumpstarting economic growth across Africa through investments in agriculture, he called on the lessons of history. The Green Revolution, he argued, "had pulled hundreds of millions of people out of poverty" and offered a roadmap for contemporary agricultural development.[12]

In the coming years, as I was developing research questions that would ultimately lead to this book, I kept recalling our visit to the Borlaug exhibit. Why did this tale about a fiercely devoted midwestern plant scientist who almost single-handedly saved millions from the ravages of starvation remain so compelling? As an American studies scholar, I think a lot about the power of some of our most enduring American stories—those about the Land of the Free, the American Dream, or a Nation of Immigrants, for example. I was coming to understand the Green Revolution story as one that held a similar kind of power. It taps into foundational narratives about the United States, but it also reproduces these narratives in unique ways. In doing so, it serves the interests of powerful American institutions, including Big Agribusiness, Big Philanthropy, and the National Security State. I became especially interested in asking how contemporary development projects on the new Green Revolution's frontier fit into this broader picture. In order to examine how these projects extend from a longer material and intellectual history, I began to build a "project archive" that pulled from a range of sources and took me from New York to Des Moines to Seattle to Nairobi (and several places in between).[13] In addition to interviewing officials directly involved in contemporary Green Revolution projects, I conducted archival research on the Revolution's history, participant-observation at the World Food Prize conference, and textual analysis of various popular media, corporate, and development institution documents. To explain this approach, I turn to a brief discussion of each method.

Archival Research

To engage with the Green Revolution's historical record, I conducted archival research on the earliest Green Revolution projects, especially the Rockefeller Foundation's Mexican Agriculture Program. The foundation had an intensive culture of record keeping that included "officer field diaries" that scientists and officials maintained in their development projects abroad.[14] Reading these reports helped me attain a better sense of how the earliest American scientists approached their work. The foundation also conducted oral histories with many key officials from its agriculture program, which proved to be particularly useful sources. These histories provided narratives about the projects that went beyond some of the bureaucratic minutiae and offered insights into the thinking of different officials. I also dug into the archives of Norman Borlaug at the University of Minnesota. As I discussed above, Borlaug's legacy as the "Father of the Green Revolution" has helped cement him as the Revolution's "brand hero."[15] He is frequently cited in development literature and the success story of the Green Revolution largely revolves around Borlaug. Reading extensively in his archives—including his oral history, various biographies and biographical portrayals written about him, and his public speeches given at academic and professional conferences—yielded insights into both the particularities of Borlaug's work and the different stories about Borlaug that have contributed to the overarching hero narrative about his influence.[16] Building upon my research on the Green Revolution's history, I turned to extensive research with the people creating its future.

Interview-Based Research

In order to understand some of the dynamics of the Green Revolution for Africa, I interviewed a range of officials involved in key projects across the continent. The Green Revolution for Africa encompasses a plethora of countries, crops, institutions, companies, and ecologies. As such, it presents challenges to write about, and even to think about, as a singular project. Thus, while I interviewed some officials about the contours of the Green Revolution in Africa more broadly, I chose to focus especially on one high-profile project: "Water Efficient Maize for Africa" (WEMA). Funded by the Gates Foundation and USAID, WEMA is a partnership between public sector scientists and the multinational agricultural biotechnology company, Monsanto, that aims to deliver drought-tolerant biotech crops to smallholder farmers across East and Southern Africa. I visited both the Monsanto headquarters in St. Louis and the offices of the Gates Foundation in Seattle, speaking to a number of representatives at both locations. On a five-week trip to Nairobi, I interviewed officials working across the partnership, especially at an organization called the African Agricultural Technology Foundation. I also interviewed public sector researchers at international agriculture research centers and Kenya's National Agriculture Research System.

I asked informants about their unique experiences with the broader project. As a public-private partnership that involves a multinational agricultural biotechnology company and multisector research institutions, the project raises questions around how knowledge and intellectual property are shared. Typically, I began interviews by asking about how research practices were shifting under the partnership model. To convey to my interviewees that I was interested in the broader context, I would often say I was interested in the "politics" of these projects. Most of my interviewees were unfamiliar with the field of American studies, so my disciplinary affiliation did not seem to signal much to them. I often asked questions about the relationship between corporate plans for biotech expansion and the role of public sector research in terms of delivering "public goods." This was—and has continued to be—a contentious subject. Because of the political nature of these issues, some of the institutions I visited had closed their doors to other researchers.[17] Yet, for the most part, the people with whom I spoke were quite willing to talk openly about their thoughts.

I was fortunate to get the kind of access to my research subjects that I had. While it is difficult to predict when and where people will make time to answer research questions, my extensive access could have benefited from a few factors. To begin, the first person I interviewed at Monsanto was not only welcoming but introduced me to several other contacts at both the Gates Foundation in Seattle and the African Agricultural Technology Foundation in Nairobi. I overheard several informants say things along the lines of "he's the guy that [Monsanto official] told us about." It seemed important that I had been vetted. In several email introductions to additional interviewees, this official recommended that people speak with me, suggesting that my "research approach [was] thorough, rigorous, well-grounded and could be a valuable academic documentation of the WEMA project at a well-regarded university in the U.S." To be clear, at no point did I ever suggest that my work might be useful for Monsanto—or anyone else I was interviewing. But the fact that at least a few of my informants thought I might do something helpful for them demonstrates the ways that research interviews are never neutral sites. Both researcher and research subject bring their own agenda.

I could have also benefitted from my positionality as a White, male researcher from a US university. In some venues, all three of these facets could have worked to my advantage. At times, aspects of my positionality might have led my interviewees to see me as a "safer" researcher with whom to speak. I also likely had more people agree to sit down to answer questions because I was, at the time, a graduate student. Many of my interviewees, after all, had conducted graduate research themselves. Even though their research was different from mine (most were scientists) many indicated a willingness to want to "help" out with my research. Relatedly, many of my interviewees had positive associations with the University of Minnesota. My home university was not just Borlaug's alma mater, but it has extensive international agricultural programs to this day. During a few conversations, interviewees would ask me if I knew a particular person in Minnesota.

Finally, I might have found more informants willing to talk because some of the organizations where I conducted interviews were hoping to improve their public image. This applied especially to Monsanto, a company with a decidedly poor public relations history related to years of controversies stemming from their production of carcinogenic chemicals such as PCBs and the active ingredient in "Agent Orange," as well as being the primary target for activists concerned with its agricultural biotechnologies, especially GMOs.[18] Based upon several of my Monsanto informants' comments, it was clear that they were very interested in sharing the "good" side of the much-maligned company with someone that might write their story along a different line than the "evil Empire" trope common among its critics.

There were, however, several times when informants treated me with a degree of caution. When one informant gathered that I was "taking a social science angle," they visibly became more reticent. And one Monsanto official wanted to make it clear that I was not using the conversation for "public communications." Another informant warned me against "burning bridges" at particular institutions. This official, who had worked for both agricultural biotechnology companies and the public sector international agricultural research centers, said that some people he knew felt "burned" by researchers. People would quit talking to me if they had the slightest suspicion that my intentions were to paint them in a bad light. Along these lines, it was clear that several of my informants were careful not to divulge much. Ultimately, though, what I was after was not really about any one person's perspective. Instead, I was trying to better understand how projects in the Green Revolution in Africa advanced through collective understandings and narratives.

Participant Observation at World Food Prize

A key venue through which I came to appreciate the power of the Green Revolution story was the site of Gates's 2009 speech: the World Food Prize Ceremony and Norman E. Borlaug International Symposium, or "Borlaug

Dialogue." More than any other event, it shows how the Green Revolution operates as a kind of living history that confirms key truths shared across the development community. It brings together hundreds of the most influential people in international agricultural development—ranging from US State Department officials to World Bank leadership to CEOs of companies like Monsanto, DuPont Pioneer, and Cargill. Many people working on contemporary Green Revolution projects in Africa also attend. Though agribusiness maintains a large presence at the conference (DuPont Pioneer, for example, has long been its largest financial backer), representatives from government, academia, and nongovernmental research and development institutions also take part. In four subsequent Octobers, I drove down to Des Moines for the conference.[19] These experiences provided the opportunity to observe the dominant discourses circulating in this community and to listen to panel discussions on topics such as public-private partnerships, digital technologies in agriculture, or communicating the benefits of agricultural biotechnologies. It was also a good place to conduct interviews. The first time I went, I met several officials working on WEMA and related projects. These initial conversations helped to open the door to several later interviews. I rely on interviews and participant observation from the World Food Prize throughout the book.

Textual Analysis of Written and Visual Materials

I complemented the above research with analysis of a range of texts related to Green Revolution projects. These included specific documents related to WEMA, promotional pamphlets and annual reports from development organizations, and media coverage of WEMA and other Green Revolution projects. Through examining different examples of how discourses around projects like WEMA circulate more widely—be it a photograph from the Bill and Melinda Gates Foundation Visitor Center or a mainstream journalism source about a Green Revolution project—I situate the Green Revolution within broader historical, cultural, and political context.

Before moving on, I want to be clear about what this book is *not* about. Occasionally when I have presented my research in public venues someone in the audience will ask me if I talked to any smallholder farmers. My answer, usually to their chagrin, is no. While ethnographic studies that consider the perspectives of smallholder farmers are undoubtedly valuable, this is not that kind of study. To be sure, I do not mean to overlook the agency of smallholder farmers—making it appear as if they have no real choice in these broader schemes. But my focus is primarily elsewhere. My approach was to "study up," to borrow anthropologist Laura Nader's phrase.[20] I was interested in the logics and narratives that drove those with the most power in these networks of agricultural development projects.

This is also not a study that homes in on the political, economic, and cultural dynamics of a particular region. There is a growing body of scholarship that offers critical assessment of Green Revolution projects in specific geographies.[21] These locally grounded accounts are crucial. This book, however, traces more macro-level connections—talking to the people that hold the power to jet-set around the continent, influence its leaders, and funnel enormous sums of money into agricultural research and development. In doing so, it tells a story about the power of the Green Revolution as narrative, one that continues to hold sway among the most powerful people in the world of global agriculture. Along these lines, each chapter offers a different lens through which to examine that story.

CHAPTER OUTLINE

Chapter 1 picks up from where this introduction began, with an account of the Green Revolution's central figure. Rather than rehash critical appraisals of Borlaug's work or assess the accuracy of claims about the Green Revolution, I describe the Borlaug narrative as a tool that shapes social memory. I show how the Borlaug story operates as a memory project that rearticulates key American myths about US exceptionalism and the country as a "nation of immigrants." In this way, the Borlaug story reproduces broader American memory projects rooted in Whiteness and settler colonialism. As such, its power lies not in what it teaches us to forget, but in how it instructs us to remember. After winning the Nobel Peace Prize in 1970, Borlaug became the Green Revolution's most prominent spokesperson. In this capacity, he would guide a growing number of people in how to remember the Green Revolution. Borlaug's lessons, however, would impart an ignorance about the historical relationships between those he described as "privileged" people in the developed world and the millions of people he viewed as desperately hungry across the Global South.

In chapter 2, I show how today's most powerful backer of the Green Revolution, Bill Gates, has taken up Borlaug's prophecy about bringing Western technologies to smallholder farmers in Africa. I examine Gates's philanthrocapitalist program for transforming Africa by focusing on the central tenets of the Gates Foundation's agricultural development: an emphasis on using philanthropy to open up new sites through which to generate profits for corporations; a narrative about Africa's "yield gap," which locates the continent's agriculture at the bottom of a global continuum of growth in terms of yields of commodity crops like maize; and a commitment to expanding the private seed sector through developing new seed companies and bringing smallholder farmers into markets for hybrid seeds (especially maize). Focusing on how the foundation promotes its efforts to the public through the words of Bill Gates and in its public-facing efforts like its visitor center in Seattle, I show how the foundation produces an approach to global poverty and hunger that at once disavows histories of racial colonialism and capitalism while maintaining racist ways of dividing people along the lines of privilege and vulnerability.

Chapter 3 turns to the material and institutional roots of today's Green Revolution by going back to where it all began: Mexico. I show how American agricultural scientists working in Mexico in the 1940s displayed what indigenous scholars describe as a possessive orientation toward Mexico's land and people. This perspective justified their efforts to collect indigenous varieties of maize from throughout the country and distribute them to US seed companies. This approach would inform maize collection efforts as the Green Revolution expanded beyond Mexico. Indigenous maize would provide the genetic backbone for American Cold War–era efforts to introduce "modern" crops to "traditional" farmers across Asia and Latin America. Projects like WEMA build directly on this legacy, as they incorporate Mexican landraces into hybrid maize bred for smallholder farmers across Africa.

In chapter 4, "Seeing Like a Seed Company," I analyze WEMA, a public-private partnership between the agricultural biotechnology company Monsanto and the world's largest public sector maize development institution. Because the project is built around introducing Monsanto's proprietary biotech traits in countries that do not have regulatory systems for agricultural biotechnology, it involves a range of "capacity building" efforts to train scientists, seed companies, and even government regulators to work with biotech crops. Drawing on over fifty interviews with WEMA officials, I analyze how these capacity-building efforts reorient public sector officials toward the profit motive. The chapter argues that because the project ties development to the expansion of private property, WEMA extend a longer lineage of colonial and developmental "improvement" logics. I situate WEMA—and the broader push to expand biotech crops across the "final frontier" of untapped markets in Africa—within a genealogy of colonial ideas about improvement and private property.

The book's last chapter, "Securitizing Smallholder Farmers on the Front Lines of the Climate Crisis," examines several novel agricultural financial technology projects that aim to manage smallholder farmers' "climate risk." I draw out connections between this budding "smallholder finance" sector and US national security policy, tracing parallels between arguments for financialized agricultural development and an emerging framework of "food security as national security." I argue that the narratives driving projects at the intersection of financialized agricultural development and national security reproduce racial hierarchies that target particular people and geographies as sites for a kind of extractive investment that financializes ever-greater pools of "climate risk."

In the conclusion, I explore key lessons gained from this book's genealogical approach to the Green Revolution. I show how the Green Revolution's emphasis on security extends to the broader context in which the US government diagnoses and plans for the climate crisis. I then outline some alternative framings through which to understand contemporary debates over the future of agricultural development and climate adaptation. Drawing on the critical tools developed throughout the book, I sketch out several frameworks that have the potential to reframe this conversation. These include: emphasizing the need to foreground issues of political, cultural, and ecological context in agricultural development; the need to consider the limitations of development projects ostensibly "African-led" that are built around the intellectual property of multinationals; the importance of questioning whether the profit motive is indeed the most robust way to generate innovative ways for tackling the issues of food insecurity and climate adaptation; a call to invite discussion of American foreign policy and US empire back into debates about the future of agriculture on a warming planet; and, finally, a reminder that history matters, and that we might take a different set of lessons from the dominant Borlaug and Green Revolution story—one that insists upon challenging enduring historical narratives that keep us from thinking relationally about global divisions of power in an era of climate crisis.

1 How We Remember the Green Revolution

During my research for this book, I took several trips to Des Moines, Iowa, to attend the week-long conference centered upon the awarding of the World Food Prize. No other name was more synonymous with the event than Norman Borlaug. A proud Iowan, Borlaug became known as the "Father of the Green Revolution" for his role in developing a high-yielding variety of wheat that, beginning in the 1960s, would catalyze dramatic increases in crop production across Asia and Latin America. He won the Nobel Peace Prize in 1970 and became the most outspoken proponent of the Green Revolution. Wanting to develop a kind of Nobel Prize for Food and Agriculture, Borlaug helped found the World Food Prize in the 1980s and was a fixture at the event until his death in 2009. In the four years I attended the conference, I saw Borlaug's likeness everywhere. Banners displaying a sepia-toned photo of the iconic scientist and the words "Norman Borlaug: The man who saved a billion lives" hang from the streetlights of downtown Des Moines (figure 1). Portraits of Borlaug adorn the walls of the beautifully restored former Des Moines Public Library Building along the river front, renamed the "Dr. Norman E. Borlaug World Food Prize Hall of Laureates." And scores of high school and college students attending the conference even sport white lapel pins that read "I [heart] Norm."

Figure 1. "Norman Borlaug: The Man Who Saved a Billion Lives," banner in downtown Des Moines, Iowa. Author photograph.

While some of this glorification of Borlaug seemed overblown, I was also struck by the sincerity with which people frequently remembered the words and actions of Iowa's most famous scientist. During the "Borlaug Dialogue" panels, speakers would remark on Borlaug's unyielding dedication, conviction to the higher cause of feeding the world, and impatience for all things bureaucratic. Their comments often sounded like recollections a family might make about a beloved ancestor: "As we all know, Norm believed in the power of science," or "as you'll remember, Norm's last words were 'take it to the farmer.'" These stories were about more than simply recounting Borlaug's accomplishments. They were about something much more deeply felt: a collective memory that conference attendees were invited to experience. Veteran development officials that worked alongside Borlaug, agribusiness CEOs who grew up learning about him, and international scientists working at US universities through US Department of Agriculture Borlaug Fellowships all shared in the sense of belonging that these memories invoked. Indeed, the story of the hardworking, though humble Midwestern scientist who "fed the world" with American ingenuity and grit was bigger than any one person.

That Borlaug has become a powerful—and useful—"brand hero" for those interested in promoting particular approaches to agricultural development has been well established.[1] Scholars have shown how the central narrative around Borlaug as a heroic scientist who prevented impending famine omits much of the historical complexities and power dynamics of the Green Revolution. And yet the Borlaug story continues to overshadow more critical appraisals of the Green Revolution's legacy, including those that consider issues of environmental damage, pesticide poisoning, wealth inequalities, rural poverty and migration, and the persistence of hunger in many of the Revolution's target countries. Clearly, more work needs to be done to reconcile the myth versus the reality of the Green Revolution. Though a worthy cause, that is not my aim in this chapter. Rather than dissect the historical inaccuracies within the dominant Borlaug story, I ask why it has remained so durable. For this task, I suggest we pay closer attention to the politics of memory in the Green Revolution.

To say that memory has a "politics" is to say that the act of remembering is inherently social. It is formed through struggles over what particular social groups remember and what they do not. In conversation with scholars who examine how collective memory is forged along contours of power like race, class, and nation, this chapter examines how the prevailing Borlaug narrative produces a way of remembering the past. When a journalist, filmmaker, or author tells Borlaug's story as one of untarnished "success," it is not simply a matter of them "forgetting" the historical record. Accounts of Borlaug continue to reproduce the same kind of historical omissions through a much more active kind of ignorance. The power of memory, as Marita Sturken reminds us, lies not in what is forgotten, but in how things are remembered.[2] To understand the persistent power of the Green Revolution we need to do more than ask whether those celebrating Borlaug know their history. Instead, we should ask how they have been taught to read the story of the Green Revolution in a way that shields them from seeing its history.

My goal here is not to level judgment on Borlaug as an individual—to declare him either a "humanitarian hero" or "menace to society," in the words of a commentary in the *Guardian*. Though much ire is generated when popular accounts diverge from the Borlaug-as-hero version of the story, we should look at the Green Revolution beyond the hero-versus-villain framework.[3] Nor am I attempting to examine Borlaug's accomplishments alongside academic critiques of the Green Revolution.[4] Instead, I take a step back from the debate over Borlaug's legacy per se and consider the power of the prevailing story about that legacy. How does the dominant narrative surrounding Borlaug teach audiences to understand the past (and present) of American international agricultural development projects? And how does it link up with broader narratives about US history? To explore these questions, I consider the Borlaug story as a social memory project—one that reproduces powerful cultural memories about American history. I begin where most accounts of Borlaug do: the prairies of Iowa.

REMEMBERING BORLAUG

Accounts of Borlaug's life trace his unwavering dedication to his work to the lessons he learned in his childhood in rural Iowa, where he was born in 1914. In a typical account, the front-page obituary from the *New York*

Times explains how Borlaug's "storied life in agriculture" began as "a boy growing up on a farm in Iowa" where "he trudged across snow-covered fields to a one-room country school, coming home almost every day to the aroma of bread baking in his mother's oven." Alongside tropes of an idyllic, yet tough life growing up in the Heartland, biographers emphasize that Borlaug was raised in a "close-knit" and "stalwart" community of Norwegian immigrants. His great grandparents came to the United States from Norway in the 1850s and biographers note that his ancestors held on to much of their Norwegian traditions as they made Iowa their new home. Yet, for Borlaug's generation, growing up in rural Iowa meant learning to see oneself not as an immigrant, but as an Iowan. This lesson was instilled at the beginning of each school day, when the children of Norwegian and Czech families would stand side-by-side and sing the "Iowa Corn Song," proudly proclaiming their belonging to the state "where the tall corn grows!" In the words of one biographer: "These immigrant children discovered in that small Iowa school that they had much in common, just as their parents found that working together to ensure sufficient food for all was more important than any ethnic or linguistic differences that might initially divide them."[5] Here Borlaug's story fits into a cultural narrative about hardworking immigrants developing a new identity as Americans through their agrarian roots. Along these lines, this same biographer calls Borlaug's life "a quintessential American success story."[6]

Borlaug's story functions as a prototypical American success story because it adheres to a powerful idea about the United States being a "nation of immigrants." Though this phrase is likely familiar to many of us, it speaks to an inaccurate version of US history. As Roxanne Dunbar-Ortiz argues, narrating American history in terms of multicultural "immigrants" becoming American erases the processes by which those "immigrants" claimed land and forged an identity as Americans.[7] This process is best understood in terms of settler colonialism, a fundamentally violent project of removing indigenous people from their native land and replacing them with a settler society. More a "process" or "structure" than a moment in history, settler colonialism is at the root of US policy, economy, and culture.[8] Historical narratives that turn "settlers" into "immigrants" teach us to remember the past in a way that whitewashes over the violence of the policies that forcibly removed indigenous people from their homelands and opened the ground for White settlers across the United States. This "multicultural approach" to US history, encapsulated by Borlaug's multiethnic schoolmates learning to declare themselves native to Iowa, works to erase the foundational violence of the United States as a settler colonial country.[9]

In a recent book, acclaimed science journalist Charles C. Mann offers an illustrative example of how popular accounts of Borlaug rehearse the settler-into-immigrant narrative.[10] Mann casts Borlaug as a technological "wizard" who applies scientific ingenuity to tackle humankind's greatest challenges. Though it offers a deeply researched biography of Borlaug, Mann's account echoes the dominant hero narrative around Borlaug and the Green Revolution. Mann certainly could have better considered critical scholarship on the Green Revolution. But rather than offer a point-by-point challenge to Mann's interpretation, I focus instead on his brief account of Borlaug's family history.

Mann recounts how Borlaug's great-grandparents, Ole and Solveig Borlaug, were some of the first "Non-Indians" to settle in their corner of Iowa, in the mid-1800s. They left Norway in 1854, after their farms had been decimated by potato blight, searching for more fertile soils in the American Midwest. Following their arrival in Wisconsin, they headed west to a "territory" that was, in Mann's words, "contested by Indians" along the Missouri River. The land to which Mann refers was, at the time, outside the formal boundaries of the United States. It was "contested," indeed. Beginning in the 1840s, the US Army increasingly built forts across the country's western frontier, from which US forces fought battles against indigenous peoples. These "Indian Wars" were integral to US westward expansion. As David Vine notes, between the 1860s and late 1890s, "the U.S. Army fought no fewer than 943 distinct engagements against Native peoples, ranging from 'skirmishes' to full-scale battles in twelve separate campaigns."[11] Mann briefly describes how the Borlaugs found themselves in the middle of an especially notable struggle within this broader war: the US-Dakota War of 1862 in Minnesota. After years of being "cheated" by state and federal governments, Mann writes, the Dakota "lashed back." "Enraged beyond measure, they killed hundreds of immigrants and defeated territorial militias in a series of battles before falling to the U.S. Army." As Mann tells it, the Borlaugs' encounter with frontier violence forced them to move back east: they "fled the slaughter" and "[drove] a covered wagon" to northeast Iowa. After summarizing the family's brush with the Indian Wars, Mann transitions to a decidedly more tranquil scene of the "close-knit Norwegian community" they found in Saude, Iowa—complete with "a general store, a feed mill, a part-time blacksmith, a cooperative creamery, and two churches." Using picturesque accounts of rural life, Mann spends the next few pages recounting Borlaug's formative years growing up in the farming community in Saude.

In emphasizing the violence the Dakota inflicted upon "immigrants" while sanitizing state and settler violence against Dakota people, Mann's account demonstrates how this period in American history is often narrated. Like many accounts of this era, Mann describes Indian violence as especially severe ("they killed hundreds of immigrants," bringing about a "slaughter"), without qualifying the violence on the part of the US Army. Mann also leaves out much context for the 1862 War, including US Army violence or how the federal government sought retribution against the Dakota by punishing thirty-eight men to death. Indeed, President Lincoln's order to execute the "Dakota 38" still stands as the largest execution in US history. This is not to say that settlers did not endure tragedies during the conflict, but that a narrative that centers indigenous violence only tells part of the story. By glossing over the wider context of the 1862 conflict, Mann's story constructs Indians as a threat to more peaceful "immigrants" like the Borlaugs.

Mann's narrative transition from the Dakota "falling" to the American military to several pages that highlight stoic, church-going immigrants working together to build lives on the Iowa grasslands also aligns with a common way of understanding American history. In this view, the era of frontier wars on "contested territory" gives way to an era of multicultural immigrants uniting under American ideals. Even if the brutality of settler violence is acknowledged, it is viewed as an unfortunate event in the past that is ultimately transcended through multicultural progress. Importantly, this way of narrating American history keeps us from grappling with the violence of settler colonialism as an ongoing project.

Mann's description of multicultural immigrant progress follows another narrative pattern common in accounts of Euro-American settlement in the United States: it emphasizes rugged, hardworking immigrants working the land while omitting how those immigrants acquired it. Mann recounts how the Borlaugs "built cabins" and grew crops, and even tells how Norman's grandfather "built up" the family's largest property, a farm of 165 acres. Yet Mann says nothing about how land was bought and sold in Iowa. We read nothing, for example, about how Iowa became part of the United States through the Louisiana Purchase in 1803, a policy that immediately put the homelands of thousands of indigenous peoples under the sovereignty of the United States. Nor does Mann mention how Iowa granted settlers, land speculators, and railroad companies large tracts of land throughout the mid-nineteenth century. This omission is noteworthy given that his story begins with the Borlaugs encountering bloody battles over "contested" land just a few hundred miles away from what would become their new home in Iowa. It would not be a stretch to say that the Borlaugs came to the United States when the country was at war over land. Yet Mann separates the politics of land on the frontier—rife with bloodshed stemming from US policies—from those in Iowa. Retreating from frontier violence, the Borlaugs take their covered wagon to Iowa, where they find "a landscape at once chillingly vacant and full of promise."[12] Immigrant communities such as those in Saude, Iowa, are seen as preordained to settle a landscape cleared of previous inhabitants.

My point is not to critique Mann for skipping over a few details, but rather to show how his book reproduces dominant frameworks through which we understand the past. The way writers like Mann ignore the politics of land and empire in relation to settler communities in the mid-1800s elucidates the power of what Keven Bruyneel calls "settler memory."[13] As a collective memory, "settler memory" functions to keep indigenous people in the background of our stories, while disavowing their relation to the politics of the present.[14] Mann's erasures illustrate the "work" that settler memory does: they teach us how to remember the past in particular ways. They center the story of heroic immigrants working together to overcome hardships, rather than the violence and dispossession of settler colonialism.

So how might this reframing of Borlaug's family history change how we remember his legacy? After all, Borlaug's family settled in Iowa nearly a century before he would plant the seeds of the Green Revolution. Does it matter whether we call them "immigrants" or "settlers"? It does—but not because it changes our view of one famous person's family history. Rather, it matters because the historical frameworks we use to engage Borlaug's family history also inform how we remember his scientific feats and his role in the projects that came to be known as the Green Revolution. In other words, historical narratives that erase the politics of land from the history of the settlement of Iowa extend to readings of the Green Revolution that erase its ties to American empire. Just as one can view the history of US westward expansion in terms of a nation "absent of empire," one can also separate the scientific feats of one man from their geopolitical context.[15]

Perhaps one of the reasons Borlaug's "quintessential American success story" remains so durable in the face of sustained critiques to the Green Revolution's empirical record is because it affirms powerful ways of remembering the past. In this sense, understanding the persistence of the Borlaug hero narrative involves asking how histories of the Green Revolution "make historical claims," and, therefore, "are necessarily making claims about memory—about what should matter to a people in their time, what is worth recalling and not."[16] Mann's version of Borlaug, like others, first tells us to remember immigrants, and not settler colonialism. It then instructs us to remember the story of Borlaug's scientific ingenuity, rather than the broader forces of US foreign policy, philanthropy, and capitalism that disseminated the fruits of Borlaug's labor. Scholars show how Borlaug lore leaves out much historical context. Yet we might extend these critiques to consider how these historical erasures are produced. To understand why commentators ignore history in their celebrations of Borlaug, we need to reckon not only with what they forget, but also with how they teach us to remember.

In many ways, the Green Revolution is still remembered in Borlaug's terms. A 2020 episode of the PBS series *American Experience* demonstrates how Borlaug's central framing of the Revolution persists, even in the face of challenges to its purported successes. The episode's filmmakers attempt to strike a fair balance. They even include voices of historians known to be critical of the Borlaug mythology. However, they uphold Borlaug's version of the story: that the Green Revolution overcame impending famine. Nowhere in the film do they explore any counterfactuals about whether widespread famine would have occurred had it not been for Borlaug's "miracle" wheat. Nor do they explore the multiple critiques of arguments about population growth and food supply.

The episode describes Borlaug's efforts to deliver higher-yielding varieties of wheat to India in terms of a solution to a "looming crisis" that "mankind would run out of arable land." It fails to account for how this "crisis" story is better understood as a narrative that held a great deal of power at the time, rather than an empirical reality about how the world works. For most audiences, these statements run the risk of being read as established fact, rather than highly contested arguments. The episode's central tension focuses upon whether Borlaug should be held accountable for some of the Green Revolution's "unintended consequences." Historian Tore Olsson manages to get an important critique across near the end of the film—insisting that food insecurity is primarily about social class rather than simply the amount of food produced. But the film is edited to minimize this and other critical moments. Instead, we get another celebratory account of Borlaug that largely upholds the prevailing narrative of the Green Revolution as a story about technological innovation overcoming impending famine. As the film draws to a close, the narrator says Borlaug "allowed the earth to bear far more people than had been thought possible."

"THE PROPHET OF WHEAT"

Borlaug's colleagues portrayed him as a driven scientist who worked long hours with a tenacity that bordered upon obsession.[17] By all accounts, Borlaug preferred to spend his days in the fields. But in the late 1960s, news that the "miracle wheat" varieties he developed in Mexico had led to record-breaking harvests in India and Pakistan brought increased public attention to the solitary scientist. As journalists began to seek him out, Borlaug began to share his conviction that Western agricultural science could catalyze profound transformations among farmers throughout what he called "underdeveloped" countries. If Borlaug's story is central to the power of the Green Revolution's core narrative, Borlaug himself would play a key role in writing that story (figure 2).

Borlaug's first television appearance demonstrates how he embraced his role as the Green Revolution's spokesperson. In 1969, the UK television

Figure 2. Norman Borlaug surrounded by trainees in Mexico. Photo credit: CIMMYT.

station ATV awarded him their "Man of the Year" prize and introduced him to the British public through a thirty-eight-minute biopic.[18] The film introduces Borlaug with a title screen that reads: "The Prophet of Wheat." Borlaug has been called a saint, a miracle-worker, and the "father of the Green Revolution," but prophet is an especially apt descriptor.[19] Though prophets are often colloquially understood as people who predict the future, the role of a prophet is better understood as a political figure that poses questions to a community, challenging them to reflect upon their collective choices—and spurring them to take action.[20] As he gained recognition, Borlaug would use his own testimony to warn a growing body of followers about the dangers of failing to advance the Green Revolution. In the process, he would solidify the Revolution's central framing as a project that raised profound moral choices between chaos and salvation.

The film opens with an eye-level, close-up shot of a sunlit Borlaug, donning his trademark short-brimmed fedora. In the film's grainy black-and-white hues, we can just see the hills surrounding Mexico's Yaqui Valley behind him. Borlaug squints into the sun and addresses the interviewer standing to the side of the camera. With his first words, he shares how he found his life's mission: "I grew up on the land, on a small farm in northeast Iowa." He continues, recalling the hardships he and his family faced. "Life was not always easy. . . . I experienced the economic depressions of the 1930s, and from this experience I felt that families on the lands—the small pieces of properties around the world—needed help from scientists and I dedicated my life to science, and especially to food production." Even at the height of his international accomplishments, Borlaug began his story with his experience growing up "on the land." Indeed, the idea that Borlaug had a lifelong connection to agriculture would become a powerful part of the story of the hero scientist.[21] Yet it was his firsthand experiences with scarcity that would provide the crux of his testimony. We learn that it was facing economic hardships that compelled Borlaug toward his dedication to serve others through science. His life experiences—his memories—of the Great Depression help to cement the authority upon which he would guide his audience toward forming their own beliefs about hunger and scarcity.

Borlaug's recollections of the Great Depression would become a key part of his larger story. His biographers cite one experience with the events of the Depression as having a profound effect upon Borlaug. Soon after arriving in Minneapolis, where he would study at the University of Minnesota, Borlaug witnessed an uprising of angry dairy farmers protesting in the streets. By all accounts, the young Borlaug was incredibly shaken by the events. The *American Experience* Borlaug episode, for example, interviews Mann in its telling of Borlaug's encounter with a "violent . . . milk riot." Mann tells viewers that the young Borlaug "saw how hunger turned men into beasts" that day in Minneapolis. Borlaug's biographers typically recreate the milk protests along these lines. Borlaug himself also recalled that day as a turning point in his life. Yet, as Tim Wise points out, Borlaug got the lesson of the "milk riot" exactly backward. The issue was not primarily about the scarcity of milk, but about how overproduction of milk had led to higher costs.[22] There was plenty of milk, just not enough money. As historians of the Great Depression have shown, though people certainly did suffer and even die from hunger during the Depression era, most of the food issues were tied to overproduction of crops, not a shortage. In well-documented cases, people were hungry because of markets, not because of a widespread food shortage.[23] Nevertheless, Borlaug would frequently reiterate this misreading in a way that fit into his worldview. To put it simply, hunger, for Borlaug, was never primarily an issue of class. He did not ask who controlled the dairy markets and why farmers and truck drivers were coordinating militant strikes against these owners (or, for that matter, why the police took the side of the bosses in the struggles in the street). The lesson of the dairy farmers' uprising in the streets is especially important because it foreshadows how a conception of scarcity as a condition of the lack of resources (rather than an issue of social class and resource distribution, as we will consider more below) gave rise to Borlaug's beliefs about hunger and poverty.

And yet viewers of the "Man of the Year" film would not have any of this critical context. Instead, they would have Borlaug's expertise as someone with direct knowledge of both agriculture and poverty (recall that the film opens with Borlaug emphasizing his "experience" as catalyzing a deeply felt conviction). This self-story, then, establishes Borlaug's purported expertise on a range of subjects outside of his direct knowledge as a wheat breeder. Indeed, Borlaug's most impassioned messages in the film would speak to general themes about history, science, and modernity. At the heart of these arguments was the idea that in most of the countries across "Asia and Africa" farming practices were "extremely inefficient." While Borlaug outlines his views on revolutionizing unproductive agriculture, the camera captures a scene in rural Mexico. Borlaug and his colleagues have gone to watch a folk band perform for the locals. We see images of a bare-chested and barefoot indigenous man, legs adorned with beaded shells, dancing in the dirt, as Borlaug explains:

> When you are asking primitive people to give up their traditional ways and their old methods, you are dealing with suspicion and traditions that are deep rooted. You must push them a bit if you are to help them. They are ultra-conservative and they are suspicious. You must make your demonstrations spectacular, so that the differences in yield of grain are tremendous. They are not differences of ten or twenty percent. They are differences of three or four or five hundred percent. So that a blind man can see them. When this is done, primitive people will be able to distinguish for himself [*sic*] how much is improved technology and how much is witchcraft.[24]

Relying, as he often would, on biblical tropes, Borlaug invokes the story of miraculous vision, in which previously unenlightened, metaphorically "blind" people come to see the truth. "Improved technology" becomes a stand-in for that which brings "primitive people" into a place of enlightened rationality. Borlaug's prescription for using the power of improved seeds to spark a psychological transformation that might convert even the most conservative peasant farmers conveyed the views around "modernization" that had taken hold of the development community around this time.[25] These ideas about modernization stem from colonial ideas about world history unfolding along a particular path, one that Western European cultures had taken first, and that non-European cultures would eventually follow. This was also a "racial historicism" in that ideas about racial superiority and inferiority informed constructions of both sides of the modern/premodern divide.[26] Scientific practices were directly tied to these racial constructions, too, as White Europeans used science as a kind of "measure of man" through which to gauge the presumed inferiority of people throughout the colonies.[27] Borlaug's narrative about primitive people extends from this longer history. In the film, it is the figure of the North American indigenous person who represents the universal "primitive people." In this way, the ideas about modernization that Borlaug lays out in the film emerge from a "settler memory" project that revolves around powerful cultural ideas about indigenous people belonging to the premodern past.[28]

Borlaug's parable about the need to shock the world's peasant farmers into embracing scientific agriculture anticipates his most vehement argument. Setting up a theme that he would return to for the rest of his career as the Green Revolution's prophet, Borlaug distills all of the world's social issues—"hunger, poverty, unrest, even war"—to one fundamental issue: food production. "The first and foremost problem in this crowded and overcrowded world is one of food," he claims. But simply producing enough food would only be half the battle. As we ride along with Borlaug in his sedan, he pulls the car to the side of the road and the camera pans out to a wide-angle shot. Borlaug has taken the filmmakers to the factory where they produce the chemical for the birth control "pill." A close-up shot from his passenger seat captures Borlaug declaring that "there are two sides to this complex human problem, the one of food production and the one of population growth." Speaking with increasing urgency, Borlaug warns: "Today the world is densely populated, and it is growing at a monstrous and frightening rate. . . . Sir Thomas Malthus predicted that we would end up in this disaster more than 150 years ago."[29]

Borlaug's doomsday warning aligned with influential "Neo-Malthusian" voices of his day that had resurrected the warnings of the British reverend and economist who declared that human population was destined to outrun the earth's food supply. Thanks to bestselling books like the Paddocks' *Famine 1975!* (1967) and Paul Ehrlich's *The Population Bomb* (1968), Malthusian thinking had gone mainstream. Ehrlich was a biologist and described Malthus's views as a kind of fundamental biological principle. Alongside Ehrlich, the Paddock brothers warned that Malthus's prediction of population outrunning food supply would soon lead to global famines. Borlaug was profoundly swept up in the neo-Malthusian craze. And while he did not always agree with their conclusions about the inevitability of future famines, he shared their interpretations of Malthus's arguments.

I will consider Borlaug's adoption of Malthus more in the following section. First, however, it is worth elaborating on the enduring power of the famous reverend's views on human population growth. Most often remembered for his arguments about population, his most influential assertion boiled down to a kind of formula that spelled out how the human population grew exponentially, while food production was tied to the earth's limited resource base. Crucially, Malthus's arguments have not stood the test of time. To be blunt, he was always wrong. Though the global human population has increased dramatically, food production has increased at least as rapidly.[30] Today, as in Malthus's day, there is more than enough food to adequately feed every human being on the planet. Nonetheless, Malthus's arguments have proven incredibly resilient. While his central claim about "the power of population" should not be viewed in terms of an established fact, it is important to consider its uptake as a discourse. To paraphrase Stuart Hall, the question of whether a discourse is "true" or "false" matters less than the one about how it becomes effective.[31]

Understanding how Malthus's argument became an influential discourse about human population demands closer attention to the context in which it was originally written. The first thing to clear up is that Malthus's most immediate concern was never an abstract "human population." He was concerned with the class politics of England. Indeed, as Nicholas Hildyard argues, Malthus's *Essay* was written as a defense of private property. In support of the movement across Britain to privatize land that was held as commons, thereby removing thousands of peasant farmers that had utilized those lands, Malthus outlined what would become a convenient argument against peasants' rights to common lands. He "furnished the privatization movement with a spuriously neutral, pragmatic set of arguments . . . that denied the shared rights of everyone, however poor, to subsistence."[32] The power of Malthus's argument proved to be its political use, reframing the debate about property relations into one about moral questions concerning the "undeserving" poor (as opposed to one of resource allocation).[33] In declaring "the power of population [to be] indefinitely greater than the power in the earth to produce subsistence for man," Malthus created a lasting morality tale about human poverty.[34] Malthus's principle of population justified poverty as the problem of the poor's insatiable appetite to consume resources—instead of a question of property distribution.

Borlaug and the neo-Malthusians he echoed were doing a similar kind of work. Emphasizing scarcity narratives about a "lack of adequate food supply," they narrowed the framework for evaluating the causes of food insecurity. Their work propelled highly politicized projects forward under the guise of a straightforward moral argument. Through his role as the Green Revolution's prophet, Borlaug became an authority in debates about hunger and poverty. Yet like the neo-Malthusians, he taught his followers to narrow the gauge by which food insecurity was measured. Drawing on his authority growing up on the land and experiencing economic hardships, Borlaug crafted a self-story that would become central to the Revolution's larger story. So doing, he contributed to the Green Revolution's social memory project, writing an enduring script of the Revolution as something akin to a parable about overcoming the dangers of unchecked population growth. Though an ardent Malthusian, Borlaug saw his work as disproving the dire predictions of Ehrlich and the Paddocks.[35] For Borlaug, the Green Revolution offered the hope of salvation to millions of people on the brink of starvation. Soon after the "Man of the Year" film aired, Borlaug would get the chance to share this prophecy from another international stage, this one in Oslo, Norway.

BORLAUG'S WHITE GAZE FROM THE NOBEL STAGE

Borlaug's legacy as the "Father of the Green Revolution" was cemented when he was awarded the Nobel Peace Prize in 1970. The day after the award ceremony in Oslo, he returned to the Nobel stage to deliver a lengthy lecture entitled "The Green Revolution, Peace, and Humanity."

Though Green Revolution projects were deeply tied to US national security interests and the United States was at that time embroiled in a protracted war in Vietnam, Borlaug did not say anything about contemporary global conflicts. Instead, he condensed the story of the Green Revolution into a narrative about Western agricultural science staving off the threat of what he called "the Population Monster." Though Borlaug asked his audience to consider humankind's future, his arguments were rooted in ideas about the past. Explaining how people in some parts of the world were "affluent" while others faced perpetual "starvation," Borlaug offered lessons about how to think historically about global poverty and hunger. From the moment he was given one of the world's most recognizable humanitarian awards, he would teach his newfound audience how to remember the Green Revolution.

Near the beginning of the speech, Borlaug told his audience that most people in their "privileged world" did not have to worry much about where their food came from or how they would feed their families. But almost half of the world's population lived in what he called the "forgotten world." Most of the people living in this separate sphere, "live in poverty, with hunger as a constant companion and fear of famine a continual menace." What led to these dire conditions? Borlaug argued that meager agricultural productivity was at the root of the so-called forgotten world's poverty. "The land is tired, worn out, depleted of plant nutrients, and often eroded [and] crop yields have been low, near starvation level, and stagnant for centuries," he summarized. However, hope remained for these impoverished souls. Increased harvests brought upon by newer varieties of staple crops like corn, wheat, and rice, were transforming the lives of poor people across the developing world. Borlaug cast the Green Revolution as nothing short of miraculous:

> For the underprivileged billions in the forgotten world, hunger has been a constant companion, and starvation has all too often lurked in the nearby shadows. To millions of these unfortunates, who have long lived in despair, the green revolution seems like a miracle that has generated new hope for the future.

Yet Borlaug cautioned that there was no time to celebrate. The Green Revolution, he stressed, had only given a small amount of "breathing space" in an intensifying "war" against hunger. A crucial obstacle, however, remained: population growth.

With characteristic fervor, Borlaug warned that unless people soon recognized "the frightening power of human reproduction," the Green Revolution's progress would prove fleeting. Without tackling population growth, increases in food production would ultimately be insufficient. In case anyone in the audience was uncertain about the intellectual roots of Borlaug's concerns, he cited Malthus by name. Yet Borlaug argued that even Malthus would have been surprised by the global picture of 1970, which he diagnosed as a "grotesque concentration of human beings into the poisoned and clangorous environment of pathologically hypertrophied megalopoles." These descriptions of overcrowded urban areas paralleled Paul Ehrlich's memorable opening passage of *The Population Bomb*, in which Ehrlich and his wife are "frightened" by the crowds they witness upon first arriving in Delhi. Like Ehrlich, Borlaug would suggest that Third World bodies have an almost inhuman character. His invocation of these teeming masses of hungry people in the underdeveloped world led directly to his central question: he asked his Oslo audience to consider whether "human beings [could] endure the strain?"[36]

Population concerns had been marginal to Borlaug's early work in Mexico, but "world population" had become the dominant framework through which global food supply issues were framed by the mid-twentieth century. Throughout the 1960s American government and development circles, a contingent of high-level advisors, National Security figures, officials from the Rockefeller and Ford Foundations, and demographers began to equate the threat of communism with the specter of population growth in the Third World.[37] In the case of India, the Ford Foundation and the US government both promoted a narrative that interpreted India's agrarian issues solely in terms of "population," rather than structural issues of unequal land distribution that had begun under British colonization. Using this framework, US officials would then interpret breakthrough wheat harvests in Pakistan and India in 1966 and 1967 largely in Malthusian terms: they had halted impending famine caused by a rapidly growing population.[38]

This subtext was just beneath the surface of Borlaug's lecture, as the fiery scientist suggested that unchecked population growth in the Global South might soon become a danger for his "privileged world" audience. In colorful language, Borlaug suggested that rapid population growth could quickly destabilize societal conditions. "Abnormal stresses and strains," he warned, "tend to accentuate man's animal instincts and provoke irrational and socially disruptive behavior among the less stable individuals in the maddening crowd." Cloaking his overblown rhetoric in the language of science, he argued that "it is a fundamental biological law that when the life of a living organism is threatened by shortage of food they tend to swarm and use violence to obtain their means of sustenance." In directing his audience to consider the animal-like urges of Third World bodies, Borlaug relies upon long-standing racist tropes that depict Black and brown people through the language of animality.[39] His biological language paralleled the broader Malthusian-infused development rhetoric of American modernization theorists at the time. As Kalpana Wilson argues, after World War II, Western governments increasingly viewed poverty and "overpopulation" as security threats. Racial imaginaries of threatening Black and brown bodies buttressed these fears. But whereas colonial representations often described Black and brown people in terms of "apathy, indolence, and fatalism, tropes which were used to justify colonial inaction in the face of famine and starvation," the development framework more often represented people in the Global South "as ominously hyperactive, incessantly 'swarming,' 'teeming' and 'seething.'"[40] Importantly, these narratives had lasting material effects. They were used to justify harsh population control programs, which Borlaug always saw as equally important to food production projects.[41]

Though Borlaug's arguments trafficked in the racist constructs of his day, it is important to "think race" here beyond simply analyzing the phrases that jump off the page as problematic. To do so, we need to consider Borlaug's Whiteness. This means more than simply talking about the color of his skin. Instead, we must engage with Whiteness as a position that is inextricable from social power. As James Baldwin memorably argued, Whiteness is a "metaphor for power."[42] Along these lines, Richard Dyer shows how Whiteness achieves its profound power through long-standing ideas about Whiteness being a kind of nonracial category—or the idea that White people are just humans, unmarked by race. "There is no more powerful position," Dyer writes, "than that of being 'just' human.

The claim to power is the claim to speak for the commonality of humanity. Raced people can't do that—they can only speak for their race. But non-raced people can, for they do not represent the interests of a race."[43] Borlaug's own claim to authority, anchored in prophetic pronouncements about "all mankind," demonstrate this relationship between Whiteness and authority. Like he would throughout his career as the Green Revolution's prophet, Borlaug's Nobel Lecture appeals to notions of universal humanity (in the hour-long speech he made eleven references to "mankind" or "all mankind"). Yet it is precisely from his position of White subjecthood, uniquely both marked and un-marked in terms of race, that he is able to do so.

Borlaug's geographical mappings—dividing the world into contrasting halves of privileged and underprivileged—also convey a way of knowing that is linked to Whiteness. As Charles W. Mills reminds us, both knowledge and its opposite, ignorance, are socially produced phenomena.[44] For Mills, "not knowing" is more than simply the absence of correct knowledge. It can also be a more active form of socially produced ignorance, in which what one claims to know, is, in fact, incorrect. Writing about the US context, Mills connects this kind of ignorance to Whiteness. For example, people socialized as White often hold inaccurate views about how deeply racism continues to influence American society. The concept of "white ignorance" points to the ways in which White people see the world through historical concepts that delimit their ability to accurately understand the reality of material inequalities as mediated by histories of racism.

Just as in our discussion of settler memory, issues of socially produced memory are important here. As Mills shows, collective memory projects also produce their inverse: "collective amnesia." Borlaug's teachings about global poverty and hunger disseminate a shared unknowing of how countries across Africa, Asia, and the Americas were *under*developed through centuries of colonization (to borrow Walter Rodney's evocative phrase from his classic book, *How Europe Underdeveloped Africa*).[45] Even as he teaches his audience to lament Third World poverty, he imparts an incapacity to understand material connections between First World "privilege" and Third World poverty. As he came to embrace his newfound role as ambassador for the Green Revolution, Borlaug would perpetuate an ignorance about the historical roots of global inequalities.

CONCLUSION: "NEW TECHNOLOGY WILL BE THEIR SALVATION"

After winning the Nobel Peace Prize, Borlaug would continue to promote the power of agricultural technologies. Some of his positions would prove controversial, such as his support of the use of the much-maligned insecticide DDT.[46] In a lecture he gave at the Food and Agricultural Organization of the United Nations in Rome, Borlaug denounced Rachel Carson, the famous biologist who catalyzed the environmental movement with her 1962 bestseller warning of the dangers of DDT, *Silent Spring*. Borlaug criticized Carson and environmental organizations like the Sierra Club for attempting to ban what he viewed as a life-saving technology. He argued that if these environmental groups were able to get pesticides banned in the United States, it would result in a tremendous loss of crops and cause food prices to quadruple. "Who then," Borlaug asked, "would provide for the food needs of the low income groups?" "Certainly not the privileged environmentalists?"[47]

In the last decade of his life, an indefatigable Borlaug would use this same line of reasoning—that privileged Westerners were keeping life-saving technology out of the hands of impoverished people in the Global South—to call for the global expansion of biotech crops. In a particularly poignant example from 2000, he wrote a polemic in the journal *Plant Physiology* entitled, "Ending World Hunger: The Promise of Biotechnology and the Threat of Antiscience Zealotry." In the article, Borlaug extolled the potential benefits of GM crops and stressed their importance in "the battle to ensure food security for hundreds of millions of miserably poor people."[48] He set up an argument that sounded quite like his Nobel lecture: global population was "mushrooming," while many of the farmers of the world lacked Western agricultural technologies that could help them overcome their poverty and hunger. But "extremists" and "antibiotechnology zealots" were largely preventing biotech crops from being used in developing countries. Lifting language from his Nobel lecture, he once again stressed the urgency of "curbing" the "frightening power of human reproduction." As the world population approached ten billion people, he argued, tools like agricultural biotechnology would be essential. Returning to the kind of "privileged world"/"forgotten world" framing he made thirty years before in Oslo, Borlaug argued: "The affluent nations can afford to adopt elitist positions and pay more for food produced by the so called natural methods; the 1 billion chronically poor and hungry people of this world cannot. New technology will be their salvation, freeing them from obsolete, low-yielding, and more costly production technology."[49] Nowhere was this mission of salvation more important than across the geographical region that Westerners most often associate with perpetual poverty and starvation: Africa.

Africa was at the forefront of Borlaug's work in his later career. Beginning in the mid-1980s, he formed a perhaps unlikely partnership with former US president Jimmy Carter and a Japanese fascist-turned-billionaire-philanthropist, Ryoichi Sasakawa, to form an agricultural development project aimed at transforming small-scale agriculture in Western Africa. Borlaug agreed to chair a new organization, called the Sasakawa Africa Association (SAA), which was founded in 1986. As president and consultant for Sasakawa's project in Africa, Borlaug traveled to the continent several times each year. Compared to the Rockefeller and Ford Foundations' programs, the SAA projects were small scale. They also lacked the kind of state backing that Green Revolution projects received in their heyday. Nonetheless, Borlaug was outspoken about the need to further expand the kind of projects SAA developed, which mostly operated through rural agricultural extension programs that introduced new varieties of seeds and fertilizers to smallholder farmers.

By the mid-1990s, Borlaug, now in his eighties, was vociferously making the argument that would soon underpin the "new" Green Revolution in Africa. The original Green Revolution, he claimed, had "bypassed" Africa. As a result, "explosive population growth" and a lack of investment in "improved agricultural technology" had increased food insecurity across much of the continent.[50] Yet when he gazed across Africa, Borlaug saw "enormous agricultural potential" waiting to be unleashed. He called for Western donors and agribusiness companies to partner with African governments in an effort "to help small-scale farmers to break out of the vicious cycle of poverty and wasted potential that they currently endure."[51] Until his dying day, Borlaug would insist that Western technologies like biotech crops could radically transform the continent's agriculture.[52] As we will see in the next chapter, his prophecy would soon be answered by a new generation of Green Revolutionaries eager to cultivate the final frontier.

2 "A Green Revolution, This Time for Africa"

Norman Borlaug is said to have died with one regret. His granddaughter, Julie Borlaug, tells how the famous American plant scientist—who had recently become only the fifth person to receive the Nobel Peace Prize, Congressional Medal of Honor, and Presidential Medal of Freedom—was consumed with thoughts of failure in his final days.[1] As the younger Borlaug recalls: "Well, he was told he had three days left to live. And he didn't speak all day. And then we finally asked what we could do. Did he want to call his family? What did he need? And he said: 'Africa. I failed Africa. I never brought a Green Revolution to Africa and I need five more years to try to do that.'"[2] Despite his family's efforts to console the dying scientist, he remained fixated on this failure. Indeed, his last conversation was allegedly about continuing a project funded by his Africa-based Foundation, which was developing a handheld tool farmers could use to measure nitrogen levels in their soil. With his last breath, Borlaug left his family with a simple instruction: "Take it to the farmer."[3]

Soon after his death in 2009, Western philanthropists, development officials, and agribusiness CEOs would invoke Borlaug's last commandment in their calls for a Green Revolution in Africa. As discussed in this book's introduction, Bill Gates claimed that Borlaug's Revolution was one of the greatest humanitarian successes of the twentieth century but that it "hadn't reached Africa." As Gates and other powerful Western donors have spent billions of dollars to transform agriculture across the continent, officials such as US Secretary of State John Kerry, World Bank President Jim Yong Kim, and the head of the African Development Bank, Akinwumi Adesina have sung Borlaug's praises while arguing that his greatest mission remains unfulfilled.[4]

That the architects of the new Green Revolution so frequently call upon Borlaug's memory matters. In telling the Borlaug story as one of untarnished success, they cement a tidy narrative about improved seeds from the West preventing impending famine in Asia. As we considered in the previous chapter, this account oversimplifies in several important ways—ignoring both the Green Revolution's geopolitical complexities and contradictory results.[5] At the same time, the prevailing narrative shores up an understanding about what a successful agricultural transformation in African fields looks like: namely, large-scale interventions to introduce higher-yielding varieties of commercial crops like maize and rice. Borlaug's story is also used to make the case that advanced agricultural technologies should drive the new Green Revolution. Though Borlaug's last words were not specifically about bringing genetically modified (GM) crops to African farmers, advocates of expanding biotech on the continent see themselves as responding to his call to "take it to the farmer." Outspoken about the potential of biotech crops to spark an agricultural revolution in Africa, Gates is perhaps the most influential person to answer Borlaug's dying wish in this way.

Gates has said that Borlaug inspired him to begin funding agricultural development.[6] The legend of the tireless, no-nonsense scientist who had little time for politics resonated with the billionaire's own brand of "impatient optimism."[7] Borlaug's arguments about shocking tradition-bound peasants with the power of high-yielding seeds also aligned with the Gates Foundation's focus on using the most innovative science and technology to catalyze development. And while Gates would mostly avoid his predecessor's Malthusian warnings about the impending threat of "swarming," hungry Third World bodies, his arguments about global poverty remain remarkably similar to Borlaug's. As I show in this chapter, Gates and his foundation also promote a view of poverty that encourages an inability to see how geographical and historical relations have produced global inequalities.

Gates's lessons about global poverty, however, reach a much wider audience than Borlaug ever could. Since the launch of the Gates Foundation in 2000, Gates has risen to prominence as a unique kind of public figure. Not only does he influence how the foundation spends its billions, but he also commands cultural capital as an authority figure in international debates about poverty, development, and climate change. During the foundation's first two decades, Gates cultivated a persona of the likable, nerdy billionaire, revered by hip-hop moguls and everyday people alike. He was as likely to appear at the World Economic Forum or the White House as on *The Ellen DeGeneres Show*, any of the late-night network talk shows, or in guest roles on popular sitcoms like *The Big Bang Theory* or *Silicon Valley*. At a time when scorn toward billionaires became mainstream (as indicated, for example, through the popularity of Vermont Senator Bernie Sanders's presidential campaigns in 2016 and 2020), media figures often avoided sharp criticism of Gates.

Yet Gates's power is about more than simply being a nicer billionaire. As the foundation has spent almost $54 billion in global health, education, and agriculture, Gates has found a growing number of ways to contribute to what Ananya Roy calls "poverty knowledge."[8] From meetings with world leaders to speeches at international conferences like the World Health Assembly or United Nations Climate Change Conference to his own writing in bestselling books and his *GatesNotes* blog, Gates advances public knowledge about the causes of and solutions to global poverty. His arguments about how to address hunger or climate change, therefore, demand critical attention not only because of his leadership at the Gates Foundation, but also because of his role as a cultural spokesperson for the power of philanthropy, more broadly. But this does not mean that we must dissect Gates's personal views. Grappling with Gates's power means thinking beyond his role as an individual. Some criticisms of Gates and the foundation veer too close to depicting Gates as a kind of evil capitalist mastermind scheming to make the rich richer. Gates's individual views, however, mean nothing apart from his Microsoft and Gates Foundation wealth. (To riff on Karl Marx, we might say that the philanthrocapitalist is merely philanthrocapital personified.)[9] Like the previous chapter's discussion of Borlaug, this chapter should not be read as a critique of Gates himself. Gates is a symptom of larger forces—and it is a better understanding of those that we are after here.

THE GATES FOUNDATION ANSWERS BORLAUG'S CALL

Gates's work to bring a Green Revolution to Africa would pick up where Borlaug left off, declaring an urgent need to get Western technologies like biotech crops into the hands of farmers in Africa. As I discussed in this book's introduction, Gates gave his first public speech about agriculture in Des Moines, Iowa, at the epicenter of the Borlaug myth-making project: the World Food Prize conference. Gates told the hundreds of corporate, government, and development officials in the audience that farmers in Africa were facing an "emergency." They were losing their maize crops to more frequent, increasingly severe droughts. The varieties they planted were not adapted to the changing conditions. If they did not get help soon, Gates explained, "millions of poor farmers" would soon be on the brink of "starvation." But Gates insisted that there was still hope for Africa's smallholder farmers. In fact, he argued, "poor farmers are not a problem to be solved—they are the solution, the best answer for a world that is fighting hunger and trying to feed a growing population."

In framing Africa's smallholders as the solution, Gates repeated what was fast becoming the central tenet of the Green Revolution in Africa: development efforts across the continent would need to focus on the needs of its tens of millions of smallholder farmers. These farmers, however, would not be able to provide the "solution" on their own. Outside interventions would need to spur a radical shift in their approach to farming—changing everything from how they purchased seeds to what kinds of fertilizer they used to how they sold their harvest. In short, smallholder farmers would need to be trained to approach farming "as a business." This mantra would soon propel much of the development projects across the continent, especially the foundation's flagship program, the Nairobi-based Alliance for a Green Revolution in Africa (AGRA) (figure 3).

The project's second director, Agnes Kalibata, would describe the continent's smallholder farming sector as a vast, untapped market. In interviews

Figure 3. AGRA offices are on the third floor of the West End Towers in Nairobi. Author photograph.

with Western journalists, Kalibata insisted that smallholders could jumpstart broader economic growth across the continent, citing a World Bank statistic that predicted that Africa's agricultural markets would be valued at $1 trillion by 2030. As Kalibata told the *New York Times*, this meant that development efforts like AGRA "need[ed] to create a value proposition around smallholder farmers for the private sector." With proper guidance, Africa's smallholders might become reliable consumers for agribusiness. AGRA would teach farmers to be more competent entrepreneurs. In the process, they would expand markets for commercial seeds, fertilizer, agrichemicals, and credit. Kalibata's assertion that what was good for smallholder farmers would also be good for the private sector encapsulates the Gates Foundation's approach to development on the continent. This formula remains unproven. Nowhere has a large-scale agricultural transformation benefited both millions of subsistence farmers and agribusinesses. Yet, for the foundation, the belief that farmers could be both successful small-scale "agropreneurs" and a customer base for corporations has become commonsense. To better understand why, we can examine the foundation's guiding principle.

"We're Capitalists Here": The Gates Foundation's Unambiguous Philanthrocapitalism

That one of the world's richest men would advocate for the power of the profit motive as an engine of positive social change is unsurprising. It is, after all, staggering profits and continued revenue from his ownership stakes in Microsoft that enables his power as a philanthropist. Gates's cheerleading for capitalism is not without qualification, however. Gates admits that capitalism's fruits are not always evenly distributed. For Gates, though, this is not a problem inherent to capitalism. Instead, he argues that poverty is simply a "market failure": markets are not in place to make something like selling medicine to poor consumers or fertilizer to small-scale rural farmers profitable for drug or agribusiness companies. The role of philanthropy, then, is to correct for these "market failures" and incentivize the private sector to move into new markets, find new sources from which to generate profits, and spread wealth around in a way that ultimately benefits society. After philanthropy has filled the market gap, its role is essentially to get out of the way and let the private sector take over. "We're always looking for an exit strategy," a foundation official explained.[10]

Because it views the profit motive as the most effective path to societal betterment, the Gates Foundation has been called "philanthrocapitalist." The foundation is capitalist not just in its aims. It also manages its philanthropic investments as if they were financial investments. Along these lines, the foundation turned to management consulting advisors from McKinsey and Company to develop rigorous guidelines for its grantees' reporting. Much like a corporation would manage its business operations, the foundation requires that organizations receiving its funds demonstrate measurable results—or "return on investment."[11] The goal of maximizing returns at every stage of its operations is central to how the foundation functions. Because of how the logic of capital runs throughout their operations, some commentators argue that the Gates Foundation has reinvented philanthropy. However, as Linsey McGoey shows, today's philanthrocapitalists are, in many ways, following the script of previous generations' mega-wealthy philanthropists like Carnegie, Rockefeller, and Ford. One way today's philanthrocapitalists are perhaps "new," McGoey points out, is in their unabashed fealty to the profit motive. Even as popular commentators question the societal benefits of the ultra-wealthy, elite philanthropists continue to command social favor through doling out their billions (or at least a small portion of them).[12]

Gates Foundation officials suggest that its capitalist orientation came straight from the top. When I visited their headquarters in Seattle, I asked a foundation official if the organization had any qualms about devoting resources to biotechnology projects that closely aligned with the business interests of companies like DuPont Pioneer or Monsanto. My informant explained that in the case of a crop like maize,

> there is a sustainable play that companies give you and that the private sector gives you. And you know, we have these meetings, these meetings with Bill Gates and some of the people who advise him and, you know, they make the joke: "we're capitalists here." And so, you know, we *are* capitalists here. And we're not confused [chuckle]. We're not the Danish AID, you know?

This quip about Gates and his advisors joking about being capitalists illustrates how, for the foundation, "capitalist" need not suggest any connotations of greed or selfishness. Rather, it conveys a shared perspective that underlies their work. This official further explained that the foundation had even stood firm when they were criticized for working with more unpopular corporations like Monsanto. "We get a lot of shit for working with Monsanto," they explained. "But there's very strong leadership from Bill Gates and Melinda about the importance of the things we're doing [and] our willingness to work with the partners who can help. And if that's Monsanto, that's terrific." The foundation's doubling down on its ties with Monsanto might raise the ire of activists. However, exposing the foundation's corporate ties should not be the end point of our critique. To fully grasp its power, we need to recognize how working closely with corporations does not betray their philanthropic interests, but rather demonstrates how they operate through the logic of philanthrocapital.

Importantly, philanthrocapitalism is not a stable thing: it "is a project, not a context," to borrow from Hannah Appel's description of capitalism.[13] This means that the philanthrocapitalist project must constantly be made and remade. From this perspective, understanding how the Green Revolution in Africa has developed along particular routes (and not others) entails greater attention to its various "making processes." We get a better idea of how the logic of philanthrocapital suffuses through today's Green Revolution by tracking its central discourses and strategies. The Green Revolution in Africa is far from a singular, uniform effort. At the same time, two related sentiments have guided projects working under its banner from the beginning: Africa's farmers are behind the rest of the world and the continent desperately needs more seed companies.

"The Yield Gap": The Untapped Potential in African Fields

If one concept unites the diverse Green Revolution in Africa projects, it is that of the "yield gap." When the Rockefeller Foundation joined the Gates Foundation to establish the Green Revolution on the continent, they claimed that the Revolution of Borlaug's era had "stopped at Africa."[14] Unreached by the Green Revolution, the story went, farmers across Africa were much less likely to grow the higher-yielding varieties of staple crops like wheat, rice, and maize that had anchored the agricultural transformations in Asia and Latin America. This led to a "gigantic gap" between the yields they could expect from their harvests and those of farmers in other regions around the world. Because of this gap, the argument went, Africa's farmers were yet to realize the full potential of their seeds and soil.

The "yield gap" has since become the primary tool for diagnosing the continent's agricultural problems. An easily distilled problem that calls for a well-established solution, the concept is pervasive across development sector, industry, and media commentaries about African agriculture. As geographer Matthew Schnurr details, the concept guides AGRA's work.[15] Moreover, coverage in the Western press has made the phrase synonymous with Green Revolution projects. According to a *New York Times* journalist, there existed an "enormous gulf between the crop yields obtained by the most successful farmers and the least successful. Farmers in the United States, for example, routinely grow five times as much corn per acre as small farmers in Africa. . . . Africa has the world's largest concentration of below-potential agriculture."[16] Gates himself repeats the "yield gap" analysis. In his 2021 book, *How to Avoid a Climate Disaster* (a number one *New York Times* bestseller), he uses a graph to represent the yield gap between farmers in the United States and farmers in Africa. Rehearsing the statistic about American farmers growing five times more corn than African farmers, Gates explains: "There's a huge gap in agriculture." "Thanks to fertilizer and other improvements, American farmers now get more corn per unit of land than ever. But African farmers' yields have barely budged."[17] Putting aside the crude comparison between farmers in one country with those across a continent of fifty-four countries for the moment, Gates's argument seems straightforward enough. Because they lack the technologies American farmers have utilized since the 1960s, farmers across Africa suffer from stagnant yields.

Yet the yield gap perspective limits the kinds of questions that are asked in agricultural development projects. As Schnurr summarizes, the "yield gap" discourse "prioritizes yield over other traits that are often more important to farmers" and "assumes a linear association between yield maximization and poverty alleviation." Altogether, the concept "privileges issues of technological enhancement over other social and political dynamics that play crucial roles in determining livelihoods."[18] Despite these limitations, the concept has become a kind of commonsense discourse, one that gains strength by repeating its own logic: if low yields are the problem, then we must do everything possible to raise yields. This viewpoint also serves as the basis for another of the Green Revolution's core arguments: Africa needs more seed companies.

"There Are Not Enough Good Seed Companies in Africa":
How Unleashing Agricultural Potential Translates
into Expanding the Private Seed Sector

In 2011, Lawrence Kent, a senior official from the Gates Foundation's agriculture program, spoke at the annual convention of the American Seed Trade Association. His message was straightforward: the "U.S. seed industry has a role to play in sub-Saharan Africa."[19] Touting the foundation's growing portfolio of grants, which at the time had already reached almost $2 billion USD, Kent explained their view that "there are not enough good seed companies in Africa." He then sketched out a broad plan to jumpstart the continent's seed sector: "We need to nurture and create more seed companies in Africa, teach farmers about the power of seed and improve the regulatory environment to make it easier for seed companies to do business in Africa." All of this would lead to a substantial expansion of the African seed industry. Kent told the American agribusiness representatives that the foundation's goals were ambitious. They wanted to "triple the amount of seed being sold in sub-Saharan Africa."

In referencing the "power of seed," Kent suggested that hybrid seeds have a nearly mythical quality. Once farmers witnessed the magic of these new seeds, everything they knew about farming would change. This recalls the kind of modernization narratives driving previous iterations of the Green Revolution. Yet, in the case of maize, hybrid seeds do, in fact, hold a unique kind of power. This is the power of seed that Jack Kloppenburg outlined in his classic book, *First the Seed*, which details how the development of hybrid corn in the 1930s precipitated rapid changes across the American seed industry. This new technology revolutionized the seed industry because hybrid seeds' dramatic increase in yield—their "hybrid vigor"—is only demonstrated for one generation.[20] Farmers are therefore compelled to purchase new varieties each growing season (instead of replanting open-pollinated varieties they have saved from previous seasons). The era of hybrid maize that began in the United States and soon spread globally was therefore also an era in which seed companies learned how to market a new kind of agricultural commodity. Today, hybrid maize is the most lucrative crop for the world's largest seed companies. (A point emphasized by one agricultural scientist I interviewed, who quipped that the industry's three most important crops are "maize, maize, and maize.")[21] But, as proponents of expanding the Green Revolution to Africa often point out, most of the continent's farmers plant open-pollinated maize varieties.[22] Realizing the kind of agricultural revolution Kent called for at the American Seed Trade Association's meeting would mean convincing millions of smallholder farmers to purchase costly hybrids each year. Bringing this transformation to fruition would, however, involve much more than simply opening new shops and stocking shelves with hybrid varieties. As Kent argued to the agribusiness officials, Africa needed more hybrid seed companies.

From its founding in 2007, AGRA would focus especially on establishing seed companies. Its first project, The Program for Africa's Seed System (PASS) has funded a range of efforts to catalyze the private seed system. In a report on its first decade (2007–2017), project officials wrote that PASS "bet boldly on Africa's private sector."[23] The report's authors note that in the postcolonial era, most African countries had national-managed seed systems, but that the era of International Monetary Fund– and World Bank–sponsored Structural Adjustment Policies had forced countries to "privatize and deregulate their economies"—a policy shift that laid the ground for further development of private seed companies.[24] In this context, PASS has focused on working with "entrepreneurs" who want to develop seed businesses. By 2017, it had incubated 114 small and medium-sized African seed companies. With funding from the Gates Foundation, the project would provide these companies with start-up funds and business consultations from seed experts who had retired from multinational corporations like Monsanto, Cargill, and Syngenta. It would also facilitate their work with small-scale agricultural supply businesses, or agro-dealers, that could market their seeds and help set up supportive regulatory environments for their companies. Hybrid maize has been a central focus of PASS companies (as of the 2017 report, 41 percent of their reported increase in production came from maize.) Though PASS has worked with a handful of farmer cooperatives and NGOs, most of their efforts have supported private companies. Their actions demonstrate how the "not enough good seed companies" sentiment has manifested in the broader Green Revolution.

Yet the Green Revolution has not only been about developing smaller seed companies. There would be a role for the big players, too. As Kent explained to American seed industry representatives, their companies might develop varieties of hybrid and genetically modified seeds with greater drought tolerance or pest resistance, which might help "convince farmers that it's worthwhile to pay" for seeds. The power of their seeds could ultimately yield millions of new customers. Around that time, outside of South Africa, very few of the continent's farmers purchased seeds from multinationals or their subsidiaries. By some accounts, the percentage of farmers that bought seeds from global companies was less than 3 percent around 2015.[25] And yet as the Green Revolution in Africa began to ramp up, signs suggested that the world's largest seed companies had "rediscovered Africa," in the words of one Gates Foundation representative.[26] Both Monsanto and DuPont Pioneer had recently made aggressive moves to purchase seed companies on the continent. Pioneer, for example, doubled its market size overnight when it acquired one of the largest African seed companies, South Africa–based Pannar, in 2013. On the heels of this acquisition, news that Western multinationals were "turning toward Africa" appeared in the Western media. Officials from Monsanto and DuPont Pioneer touted their companies' long-term visions for expansion on the continent to *Bloomberg Businessweek* in 2016, for example.[27] DuPont Pioneer's president, Paul Schickler, argued that climate change brought unique challenges that multinational corporations were well-prepared to address. The *Bloomberg* journalist summed up Schickler's perspective: "The patience of early investors [on the continent] will be rewarded." While at the turn of the century one could argue that the multinational companies were not paying much attention to African markets, the tone has decidedly shifted toward viewing the continent's markets as an increasingly viable opportunity.

A wave of high-profile mergers and acquisitions would rock the global agribusiness sector between 2015 and 2017. These led to even further consolidation in seeds and agri-chemicals in particular, sectors with already high levels of concentration. The "big six" global agri-seed-chemical companies are now "the big four."[28] All signs point to now even larger multinationals continuing to set their sights toward expanding markets across Africa. Corteva Agriscience, the agricultural company that was spun out of the merger of Dow and DuPont, has staked out bold plans for expansion on the continent. It opened the doors to a new regional hub for Eastern and Southern Africa in Nairobi in 2019, promising to "completely disrupt [the region's] agricultural sector."[29] The mega-mergers heightened the level of public attention directed toward a sector in which mergers and acquisitions had been increasingly common. Like many of the world's most sizeable corporations, large companies like Corteva and Bayer have long lists of subsidiaries, some operating under the brands of their parent companies and others under different brands. As researcher Phillip Howard has detailed, the history of the U.S. seed, biotechnology, and agrichemical industries has been marked by intensifying corporate consolidation.[30] As the multinationals expand their footprint on the continent, questions emerge about whether their long-term interests include acquiring the most profitable seed and input companies, including those developed through Green Revolution projects like PASS.[31]

The possibility of future corporate consolidation is a key concern of farmer and activist organizations on the continent, who worry that increased corporate control of seeds and inputs will make farmers more dependent upon the whims of global markets and, ultimately, give them less control over their own farms. But when I asked officials from the Gates Foundation about the prospect of the largest multinationals buying up seed companies in Africa, they would answer that these kinds of takeovers were inevitable. A senior Gates Foundation representative talked about how the multinationals might eventually buy up the companies AGRA was starting. I had asked if recent high-profile mergers and acquisitions had changed anything about the foundation's approach. The official explained that mergers and acquisitions were surely on the horizon for some of the 114 seed companies AGRA had created. "In the early stages, to reach smallholder farmers, you need as many players as possible. But, of course, as the industry grows and matures, you will start to see some mergers and acquisitions. Later. But that's healthy."[32] They went on to detail what they viewed as the most likely scenario for what mergers and acquisitions would look like down the road. Companies would buy up other companies but leave their brand in place and use the company as a "conduit to drive the pipeline." I wanted to press the foundation on this issue of working to benefit the bottom line of the biotech behemoths. I asked if they thought some of the AGRA companies would end up being subsidiaries of the multinationals. "Of course, of course, which is a good thing because they become part of something bigger. There is more capital. They can expand. Reach more farmers, which is healthy."

I asked if the Gates Foundation would be concerned with this process. "No," the foundation official answered. "No, it's a good thing."

If the logic of capital hinges on the remarkable power by which money generates more money and value generates more value, then the logic of philanthrocapital is to invest in a way that introduces this process where it has not fully taken hold.[33] It should come as no surprise, then, that the Gates Foundation official articulated their objectives in terms of growth leading to "more capital." Because of capital's ceaseless pursuit to become "something bigger," seed companies operating under its logic must expand. Industry consolidation, from this perspective, is not something to be avoided, but is actually "healthy." My question hinted at a kind of moral judgment ("aren't you really just doing the bidding of a *greedy* agribusiness?"). But my informant's candid dismissal displays how the foundation should not be thought of as generous or greedy. Rather, operating under the logic of philanthrocapital, they must invest with the explicit aim of generating "more capital." Asking ethical questions about right and wrong thus keeps us from grasping the foundation's function: expanding social relations in which the logic of capital becomes the engine for social change. A smallholder farmer becoming a consumer of seeds and agricultural inputs that will generate revenue for multinationals is not an aberration. It is the logic working as it should.

The Philanthrocapitalist Gaze

These points that Gates Foundation officials outline—the yield gap, the emphasis on strengthening the private sector, and the vision that more capital, even in the interest of global multinationals is the solution, not the problem—all point to the way that food insecurity across Africa is increasingly viewed from a philanthrocapitalist gaze. As with generative concepts like the "male gaze" or "White gaze," the philanthrocapitalist gaze simplifies what it sees as it fails to consider how its own perception is constructed. It does not recognize the material conditions that underpin its own privileged perspective. (Of course, the philanthrocapitalist gaze is often also a masculine, White gaze as well.) Much like Borlaug's depictions of the "privileged" and "forgotten" worlds, the foundation's philanthrocapitalist gaze reproduces a kind of ignorance about the historical conditions that have produced those kinds of harsh separations between global haves and have nots. In other words, the philanthrocapitalist gaze perpetuates what Mills calls a "global white ignorance": a worldview committed to Whiteness that erases past racial injustices and insists that global inequalities can be remedied without addressing legacies of racial exploitation.[34] Extended to global geographies and histories, the concept of "white ignorance" clarifies how the history of European colonialism as a racial project that systematically looted the continent of Africa is whitewashed or minimized in conversations about contemporary development. For Mills, this sanitized past keeps us from asking how contemporary wealth in the North is built upon legacies of "racial exploitation of the labor, land, and techno-cultural contributions of people of color."[35] Along these lines, when philanthrocapitalist projects look at agriculture solely in terms of yields, seed companies, and capital, they keep us from asking deeper questions about history and power.

Consider, for example, the "yield gap" formulation that Gates and others routinely use to frame the problem of food insecurity in Africa. From this perspective, African farmers are understood to be advancing along a linear trajectory, in which the end goal is to adopt the input-heavy agricultural practices of farmers in the American Corn Belt. Questions of context and what methods work at different scales are left aside. At the same time, the "yield gap" perspective locates the source of the problem on African soil, rather than considering how land, seed companies, and agricultural products across the continent have been *under*developed. As one example, historian James McCann details how the maize seed industry across Southern and Eastern Africa grew out of the colonial system, which privileged White, settler growers with larger landholdings at the expense of numerous smallholder farmers.[36] This led to a two-tiered system in which hybrid maize varieties were developed to thrive under the conditions of larger, settler-owned farms and open pollinated varieties became the seeds of choice for the region's millions of smallholder farmers. Even after the end of formal colonialism, the historical legacy from these divisions remains. While there are sharp contrasts between the levels of yields farmers in Africa typically harvest and those that farmers in the United States harvest, telling the story solely in terms of the "yield gap" fails to account for how histories of uneven development influence contemporary conditions—at scales spanning from the largest landholders in a particular region to complex global commodity chains controlled by multinational corporations.

Gates has become a powerful storyteller for contemporary capitalism.[37] Yet we should consider how these storytelling practices—insisting on the power of markets, techno-optimism, and arguing for the power of unleashed capitalism—are rooted in a deeper kind of ignorance. Gates's solutions to the "emergency" of food insecurity and climate change across Africa fosters a perspective that cannot recognize how these issues extend from centuries of colonial and neocolonial projects centered upon violent extractions of the continent's wealth of "resources"—plants, minerals, and, especially in the case of the history of transcontinental slavery, people. Yet just as with our discussion of Borlaug's failure to account for the relationship between the two worlds in his Nobel Lecture, the Gates Foundation advances a worldview that is fundamentally ahistorical. Even more so than the father of the Green Revolution, however, Gates's perspectives have wide ranging political and cultural appeal. His views also inform decisions about how massive sums of philanthrocapital will be invested. The philanthrocapitalist promise to address poverty, hunger, and climate change adaptation through expanding the reach of corporate agribusiness generates durable, widespread narratives about the significance of these intertwined issues. As I outlined in this chapter's introduction, the Gates Foundation has tremendous reach not only in direct development work, but also as a broader cultural force producing "poverty knowledge" in multiple ways.[38] This knowledge extends beyond the Green Revolution and contributes to wider conversations about poverty, global health, and climate change adaptation. In this context, thinking more critically about what the foundation teaches us to see—and not see—is imperative. To get a clearer sense of how the foundation imparts lessons about global poverty, this chapter's final section turns to a firsthand account of my own trip to its headquarters in Seattle.

"A GREEN REVOLUTION, THIS TIME FOR AFRICA"

THE GATES FOUNDATION'S PEDAGOGIES OF POVERTY

Though Gates's brand of techno-optimism has received a fair share of criticism in both scholarly and mainstream commentaries, less has been noted about how the Gates Foundation promotes its understanding of global poverty.[39] To better understand how the foundation teaches public audiences about poverty, I visited the Gates Foundation's Visitor Center—called the "Discovery Center"—at their headquarters in Seattle (figure 4). I had made the trip to Seattle to spend a few days interviewing Gates Foundation officials about projects like AGRA and Water Efficient Maize for Africa. But I also wanted to see how the organization presents itself to the public—how it builds its brand around particular ideas, imagery, language, and employee culture.

The foundation's $500 million complex is architecturally striking and sits in the heart of the city's tourist district, just across the street from the iconic Space Needle and a five-minute walk to popular destinations like the Museum of Pop Culture and the Children's Museum. I had passed the Discovery Center on my way to the main campus and had noticed several prominent representations of their agricultural development programs, including an infographic about AGRA and a curious sidewalk sculpture consisting of a row of concrete grain sacks. Eager to learn more about how the Discovery Center teaches visitors to the Gates Foundation how to understand their own relation to global poverty, I decided to spend a few hours walking through its galleries on the last afternoon of my visit.

The first thing I notice after walking through the center's front doors is a wall of words seemingly floating behind the information desk. Made up of layers of plywood letters stacked together, the wood block quote juts out from a plane of clear glass: "Whatever the conditions of people's lives, wherever they live, however they live, they share the same hopes, the same dreams as you and I—Melinda French Gates." This appeal to universal humanity expresses an ideal at the heart of the Gateses' philanthropic brand. The foundation's mission statement (displayed in another oversized block quote on the visitor center's concrete facade) also invokes universality: "Every person deserves the chance to live a healthy, productive life." I had seen similar platitudes about global humanity in the artwork and signage throughout the foundation's campus, so Melinda's quote

Figure 4. Gates Foundation Discovery Center, Seattle, Washington. Author photograph.

was an unsurprising epigraph for the Visitor Center. But I was struck by its second-person point of view—by the way Melinda's "you" directly appealed to me.

I soon discovered this mode of speaking directly to its visitors was central to the Discovery Center's exhibits. As I made my way through its high-ceilinged galleries, I encountered interactive displays about philanthropy and poverty. I came across a wooden desk with two keyboards and a placard reading: "what would your foundation do?" The ever-expanding scroll of past guests' answers was projected on a bank of monitors on the wall. Across the room, an exhibit dedicated to the foundation's positions on more controversial issues like vaccines and GMOs also solicited my input. After watching brief video messages from the Gateses and Foundation CEO, Jeff Raikes, I could enter my own thoughts into a similar digital archive. The placard discloses: "We know that not everyone supports our methods. . . . Even when we disagree, we applaud everyone working to help find solutions to big problems. We encourage you to get informed and join these important discussions." Even the drinking fountains offer an object lesson. While quenching my thirst I notice an uncaptioned photo of a woman walking across what appears to be quite dry earth carrying what I assume is a vessel of water on her head. Bold text on the white wall above the drinking fountain asks: "What if you had to walk 3 miles for this water?" After learning about the foundation's global philanthropy, I get a chance to share my own ideas about making the world a better place. A floor-to-ceiling "share your cause" armature features hanging rows of postcards where guests have left inspirational messages ("I'm going to get out and . . ."; "I support . . . and you should too"; "I volunteer for . . ."). I'm asked to take one and leave one.

On the floor surrounding this "share your cause tree," I notice a trail of footprints painted onto the wood floor that extends along the main corridor, back down toward the entrance. The dark brown footprints contrast sharply with the lighter hues of the reclaimed ash hardwood floor. Intrigued, I follow the trail to the other end where I find two metal buckets, one labeled "16 lbs." the other "2 gallons." Eager to try out my strength, I lift the buckets, bringing into view an eye-level display with a photo of three young girls. It explains that women and children all over the world have to walk miles just to get access to water. Again, the second-person address invites empathy for these poor girls: "Could you carry water for your family?"

As I would later learn, the Discovery Center was designed with the explicit goal of prompting visitors to reflect upon their own sense of identity through encountering lessons about global poverty. In an interview shortly after the center opened, its curator spoke about how they strove to design exhibits that would create a "very personal" experience, through which visitors could make connections between the foundation's work and "their own lives."[40] Though the center's interactive exhibits invited me to imagine myself as having something in common with the global poor, they did so through emphasizing that these people lived lives much different from mine. This message was reinforced through scenes showing intimate details of people's lives: close captured portraits, large photographs of women weighing their children at open-air clinics, and, in the restrooms, photographs on the stall doors of pit latrines from around the world. All of these reinforce the materiality of poverty. Each image invites an assumed Western guest to imagine what their own lives would be like under such drastic material differences.

While we are asked to share the Gateses' values about "all humans," the center's repeated object lessons reinforce that some categories of humans experience profoundly different realities than "us." In the words of Kaiama Glover, the exhibits "bring us and them into a place of temporary and hierarchized false intimacy in which categories of human beings are demarcated for all the world to see."[41] Meanwhile, the Discovery Center's visual texts work through a contradictory impulse that is at once inclusive and exclusive, appealing to universal humanity while "staging" stark differences between categories of human—between the assumed "us" of Melinda's epigraph and the "them" depicted in decontextualized images as capital "O" Others.

By taking in these intimate scenes of difference—toilets, people fetching water or taking babies to clinics, bare hands cupping maize seeds—visitors are invited to both gaze upon and imagine racialized bodies. Many of the images include women and children that they would recognize as Black and/or African. As Glover reminds us, humanitarian discourses have long relied on associations between Blackness, materiality, and abject poverty.[42] The Visitor Center's staged ethical encounters reproduce these kinds of associations. In the "Walking with Water" exhibit, that picture of three young girls above the water buckets prompts me to reflect on the difficulty of their everyday lives. As I lift the buckets, their thin metal handles dig into my hands and my shoulder muscles soon burn. I feel empathy toward these girls; yet, as their picture drops back into the exhibit, I also imagine how different their lives are from mine. Race provides the grammar through which I imagine these material and bodily differences—and through which they become naturalized as something that is commonsense. These lessons reinforce powerful ideas about racial difference, in which White, Western audiences come to associate Black and brown bodies in the Global South in terms of being naturally poor. These lessons recall the pedagogy of the widely popular American magazine *National Geographic*, which taught a largely middle-class Western audience to understand material difference as racial difference.[43] The Discovery Center's explicit instructions about tangible, everyday conditions of impoverished people across the world similarly naturalize a way of thinking about global inequities through bodily, racialized perspectives.

Though race was central to how the center's object lessons functioned, the stories the foundation told across its exhibits elide questions of race. The captions posted alongside intimate scenes of difference prompt viewers to read these images of bodies through other lenses, especially those of technology and access to markets. Yet this contrast is not because race was an oversight, simply not thought about by the curators. Instead, this contradiction illuminates how the center operates through and reproduces "postracial" logics. The Gates Discovery Center teaches us to "see" race, but to do so through vantages that disavow structural issues of race and global poverty. In teaching us to consider our own individual actions vis-à-vis the subject of development in the Gateses' philanthropy, the Discovery Center operates through a postracial understanding of poverty. This perspective disentangles the everyday experiences of the world's poorest people from centuries-long histories of capitalism's uneven development. Instead, we are encouraged to view the Poor Others in the photographs as potential entrepreneurial subjects capable of investing in their own lives.[44] The visual narratives taught through images of Black and brown people reinforce difference, while teaching us to understand that difference, in Roopali Mukherjee's words "as rational rather than racial."[45] Directed toward an assumed Western subject, the center's second-person narratives solidify postracial imaginaries about global poverty by emphasizing the individual agency of Western subjects and technological solutions to global poverty. We can see here how the visitor center operates along two distinctly postracial lines: First, its narrative of universal humanity centers the agency of Whiteness.[46] And, second, through framing poverty as an issue of individual choice and market access, it disavows structural questions about the centrality of race to capitalism.

Over eighty thousand people, many of them K–12 students, visit the Discovery Center each year, making it function as a site for public outreach about the foundation's approach to philanthropy.[47] It also teaches its thousands of guests to think about poverty through particular lenses. Through the public platform their foundation enabled, the Gateses became prominent storytellers for contemporary capitalism. In Nicole Aschoff's terms, they operate as "prophets of capitalism": acknowledging that there are problems within capitalism, but, through a discourse of techno-fixes and innovation, they maintain that capitalism can be fixed to better serve the needs of everyone. The Gateses are clearly powerful storytellers for capitalism. But we should also attend to how their philanthrocapitalism serves as a conduit for postracial thinking. The foundation's focus on market-based solutions to global poverty consistently reproduces a de-racialized understanding of inequality, even as it associates material poverty with people and places that are racialized as nonwhite. Borlaug's more overtly racialized language of threatening, animal-like bodies has been replaced with a postracial framing that disavows race as a modality of power even as it buttresses racialized constructions of difference.

While much work on postrace has been applied to specifically US contexts, the concept can also be used to see how enduring American ways of thinking about and enacting race are reproduced through philanthrocapitalist projects outside the borders of the United States.[48] There are certainly numerous intersections of power occurring through lines of race, ethnicity, class, gender, and nation across different scales of agricultural systems in Africa. My point, it is worth clarifying, is not to subsume the entire Green Revolution as a homogenized project that operates under one guiding postracial logic. Nevertheless, the outsized power of the Gates Foundation to write narratives about the Green Revolution in Africa as it designs so much of its on-the-ground operations demands that we pay close attention to how it conveys issues of race, history, and power.

CONCLUSION: WHAT THE PHILANTHROCAPITALIST CANNOT SEE

This chapter has examined the narratives that guide the Gates Foundation's efforts to bring a Green Revolution to Africa. This is not because the foundation is the sole catalyst for all the agricultural development projects that have cropped up across Africa since the early 2000s. Multiple state, corporate, and nongovernmental institutions shape the continent's ever-changing policy and project landscape. At the same time, the Gates Foundation has played a leading role in determining what questions are asked and what projects are pursued. Since it began funding agricultural development work in 2006, the foundation has fast become the most influential donor in international agricultural development. It has funneled billions of dollars toward incubating seed companies, training scientists, supporting university agricultural science, funding research and development, training regulators, advocating for policy changes, training journalists, and developing partnerships with NGOs, private companies, and governments. Aside from its strictly financial clout, the foundation has also shaped the broader discourse about agricultural development in Africa. Its categories of analysis and terms of debate shape the agricultural development conversation at multiple levels, from national-level researchers pursuing projects to address yield gaps to international conferences charting the course for future corporate and public investments. The foundation's power also extends beyond the confines of its grant-making and public relations programs. Indeed, it is difficult to draw a sharp line around where the work of the foundation begins and ends. For example, former foundation officials have moved into prominent jobs in government agencies (notably, an early leader of the foundation's Global Health and Agriculture programs, Raj Shah, went on to lead USAID during the Obama administration, where he transformed the agency to further prioritize partnerships with the private sector).[49] The foundation also continues to solidify partnerships with a range of international institutions, ranging from the

FAO to the World Food Organization to USAID. Through these forums, the foundation is an increasingly powerful voice in shaping international debates. It also contributes to high-level conversations about how to address climate change.

The kinds of lessons Gates offers in his book *How to Avoid a Climate Disaster* follow the philanthrocapitalist script. Commentators have argued that the book offers little in terms of examining questions about power and justice.[50] This chapter extends that critique by showing how Gates and the foundation advance a de-politicized account of the causes of the climate crisis and the solutions that might mitigate it. For Gates, addressing climate change is about advancing the right technological solutions, rather than an issue that demands tough questions about responsibility, reparations, and justice. This approach to climate change does not ask how global divides between rich and poor have been mutually constituted. Gates refuses to consider how climate vulnerabilities are distributed unevenly precisely because of longer histories of racial colonialism and capitalism. He fails to "see race" during climate crisis.[51] Like his Green Revolution predecessor Borlaug, Gates teaches audiences how to understand the past. In calling for more hybrid seeds, more seed companies, and more capital to transform farming across Africa, Gates cultivates a way of thinking about the causes of food insecurity that disconnects the North from the South. His philanthrocapitalist gaze "distances and confines" the causes of food insecurity within continental borders, rather than addressing the living history of centuries of colonialist and neo-colonialist pillaging of the continent.[52] While Gates and other leaders of the Green Revolution in Africa tout the philanthrocapitalist promise, more attention to their guiding principles would lead to a different set of questions about how to address very real inequalities in terms of hunger and climate adaptation. To ask these questions, we need to historicize the material and conceptual lineage of the Green Revolution. With this imperative in mind, the next chapter traces the seeds of today's efforts to introduce hybrid maize in Eastern Africa to the first Green Revolution's roots in Mexico in the 1940s.

3 "The Landraces Are in the Hybrids"

While interviewing a scientist from the International Maize and Wheat Improvement Center (known by its Spanish acronym CIMMYT) in Nairobi, Kenya, I was struck by a display of dried cobs of maize framed in a shadowbox on the wall. The colorful cobs stood out among the otherwise ubiquitous office furnishings—a whiteboard, banks of fluorescent lights, a wooden conference table stacked with papers, and pleather swivel chairs. Locked under the box's glass cover and matted against a black backdrop were three rows of variously sized maize varieties. The gold nameplate at the bottom of the box read "Mexican Maize Landraces"—the varieties of maize recognized as indigenous to Mexico. Each of the twenty-eight cobs featured a similar tag with a taxonomic name like "Bolita," "Olotillo," and "Reventador." This prominent display of indigenous maize from Mexico in CIMMYT's East Africa headquarters was, at first glance, curious. I was in Nairobi to research the workings of one of the flagship projects of the Green Revolution in Africa, "Water Efficient Maize for Africa." Funded by the Bill and Melinda Gates Foundation and the US Agency for International Development (USAID), the project brought together CIMMYT and the multinational agricultural biotechnology company Monsanto in an effort to develop drought-tolerant, genetically modified maize, reform regulatory systems, and build the private seed sector in East and Southern Africa.[1] My conversations with WEMA officials had mostly been about the political and ecological context of maize farming in Africa. But that display of the landraces of maize also offered an object lesson about the roots of the Green Revolution in Africa.

Today's large-scale efforts to transform how maize is grown across Africa can be traced through CIMMYT's institutional history, back to the Rockefeller Foundation–sponsored Mexican Agriculture Program (MAP) of the 1940s and 1950s. It was through the MAP that Norman Borlaug bred the wheat varieties that catalyzed drastic increases in yields across Southeast Asia. Though overshadowed by the legend of Borlaug, the program's expansive work on Mexico's most widely cultivated crop—maize—would also yield effects that continue to be felt far beyond Mexico. From the MAP's modest beginning, when a handful of earnest American agriculturalists partnered with Mexico's Ministry of Agriculture to transform the country's agriculture, maize was at the forefront of their efforts. Unlike the United States, where maize farmers were rapidly industrializing around the newly developed hybrid seed technology, almost all of Mexico's maize growers farmed on small plots of land with maize plants they had sown for generations. Many of these campesinos farmed on collectively owned land granted to them through Mexico's revolutionary land reforms that began in 1917 and reached a high point in the 1930s. Though the country's political, cultural, and ecological landscape seemed a world away from their own, the MAP scientists were eager to apply their knowledge to improve upon what they saw as vastly untapped potential held within the country's maize crops.

This improvement mission was also a scientific expedition. Indigenous people first domesticated maize from an ancient grass thousands of years ago in what is now Mexico. Recognized as the crop's "center of origin," the country holds unparalleled genetic diversity. Eager to mine these genetic riches for undiscovered beneficial traits, the MAP scientists conducted expansive "collecting" trips throughout the country. From the start of the project, they would systematically collect and catalog maize varieties from throughout Mexico and Central America—and distribute those samples to American seed companies and international development institutions across the global South. Though they have been cross-bred with different varieties over the years, the hybrid maize varieties that CIMMYT distributes today—as well as those private seed companies sell across the world's tropical areas—hold the genetic traces of the seeds collected from those early expeditions. A CIMMYT official who had worked with seed companies across Latin America and Africa emphasized this lineage, reminding me that "the landraces are *in* the hybrids."[2] This genealogy of maize certainly matters for how we think about the pedigree of contemporary hybrids, whether they are publicly available through research organizations like CIMMYT or the tightly controlled intellectual property of corporate multinationals. The story I tell here, though, is not only about this material lineage. It is also about the ideas that guided the scientific efforts to extract the genetic wealth held in the landraces of maize—and how those ideas came to be embedded in the seeds of the Green Revolution.

The American scientists the Rockefeller Foundation first sent to Mexico displayed what indigenous scholars describe as a "possessive" orientation toward Mexico's indigenous people, their land, and cultural heritage—including seeds. Rooted in the US settler colonial project, this logic viewed indigenous people and their maize in terms of the ancestral past of modern, developed agricultural science. Even as they claimed the genetic resources that indigenous people had cultivated in their locally adapted maize varieties, the earliest Green Revolutionaries distinguished their efforts as a decisive break from the cultural practices of indigenous peoples. As they extracted seeds from fields and villages across Mexico, they reproduced this possessive logic. As they began to share the varieties internationally, this logic would underpin subsequent maize breeding projects along an ever-expanding agrarian frontier.

This chapter retells the origin story of the Green Revolution. Known mostly as the center of Borlaug's wheat development efforts, the program is widely lauded in mainstream accounts of the Green Revolution. Scholars of the Green Revolution also critique the program for directing Mexico's agricultural policies away from political concerns such as land reform and toward the benefit of larger, capitalist farmers.[3] Yet neither critical nor celebratory accounts have adequately considered how the project's scientists contrasted their work with the agricultural practices of Mexico's Indians, which is precisely where this chapter begins.

SURVEYING MEXICO'S POTENTIAL

That the Rockefeller Foundation pursued its first agricultural development program south of the US border owes much to an iconic American corn breeder. Henry A. Wallace had perhaps done more than any other person to advance the development of hybrid corn in the United States. Wallace was an early adopter of the breeding practices that led to increased yields or "hybrid vigor." He founded the Hi-bred Corn Company in 1926, which later became the Pioneer Hi-bred Corn Company in 1935. The company was at the center of a swift transition in American agriculture, in which most corn growers went from planting open-pollinated varieties to purchasing seed each year from companies like Pioneer. After serving as secretary of agriculture in the 1930s, Wallace became Franklin D. Roosevelt's vice president. The United States was working to develop a "friendly neighbor" policy with Mexico as World War II expanded across Europe, so FDR sent Wallace to Mexico to attend the inauguration of Mexico's incoming president, Manuel Ávila Camacho in 1941.

According to the usual telling, Wallace stayed in Mexico for a few weeks as the guest of the US ambassador. He was interested in the country's corn industry and worked with officials from Mexico's ministry of agriculture and the US Department of Agriculture to begin testing American hybrid maize in Mexican soils. After returning from Mexico, Wallace met with Rockefeller Foundation leadership and urged them to consider how they might assist in improving Mexico's "inefficient and even primitive" agriculture.[4] At the behest of Wallace, the foundation hired a team of prominent American agricultural scientists to drive south of the border and spend a summer surveying Mexico's agriculture. Richard Bradfield, a Cornell University soil scientist, Paul Mangelsdorf, a Harvard University plant geneticist, and E. C. Stakman, a plant pathologist from the University of Minnesota formed the "Survey Commission." They spent two months driving across the country, putting over five thousand miles on their GMC Suburban Carryall (figure 5).[5]

In the report they sent to Rockefeller Foundation headquarters in New York, the commission declared that Mexico had "many of the aspects of an overpopulated land."[6] Yet the commission argued that simply cutting down "jungles" and turning more hectares into farmland could not improve Mexico's agriculture. Even where they saw efforts to clear "raw land,"

Figure 5. Members of the Rockefeller Foundation's Agricultural Survey Commission to Mexico, 1941. Photo credit: Rockefeller Archive Center.

they maintained that the fundamental issue was that the land was inherently poor quality.[7] In their understanding of Mexico's "land problem," the commission drew upon the ideas of the influential Mexican intellectual Daniel Cosío Villegas, who argued that Mexico's economic poverty was a result of its "natural poverty."[8]

The commission's representations of Mexico's "natural" conditions extended to "cultural" conditions as well. To get a sense of the rural culture, the Americans hired locals to take them on the backs of trucks or donkeys into areas beyond the highways. No one on the team spoke Spanish, so they brought a recent PhD in botany from Harvard, who had worked in Mexico and knew Spanish. Still, because many of the people they encountered in their journey spoke indigenous languages, they had difficulty communicating. They later elaborated on these backcountry trips, writing:

[We] learned to appreciate some of the problems and the hopes of the humbler peoples who lived near the end of the trail, close to the land but far from water in the drier areas and close to the water but too far from dry land in the wetter areas. And the horizon was too close to the earth for many people in all areas, because their land was poor, tillage was poor, and they were poor.[9]

The Survey Commission's report makes it clear that they identified many of these triply poor people as Indians. They described "large populations" of Indians that lacked agricultural skills and "economic resources."[10] And the report concluded: "the basic Indian nature of the population is a fact of paramount importance." Extending their blanket assessment of Mexico's "natural poverty" to the country's indigenous population, the commission conflated economic poverty with indigeneity. Its views of Mexico's population as largely poor and mostly Indian influenced how members of the commission viewed Mexico as a country steeped in the "old" ways, but emerging toward modernity.

After traveling across the country, the commission sent a hefty report back to New York that categorically detailed the stages of this progression. "Mexico," they concluded, "was a land of violent contrasts." As evidence of this assessment, they included 143 photographs of Mexico's crops, livestock, universities, and people. Pictures of farmers juxtaposed what the commission viewed as modern and traditional worlds: "A tractor with a gang of steel plows may be encountered in a field adjacent to one tilled with a crude wooden plow." Similarly, the photos depicted colleges across the country, with captions noting if they were "modern" or still "in the old tradition," often through direct comparisons to American universities. They also showed various maize plants across different regions, noting "harsh contrasts" between plants that were "knee high" to those over "20 feet high." Despite being impressed with some of Mexico's maize, the commission reported that "much of the corn [was] grown on land which is not suited for its production."

The photos the commission used to represent Mexico's "contrasts" also convey the scientists' keen interest in improving plants and animals through the scientific management of breeding. They described visiting a few "modern" dairies, including one that owned a "pure-bred" bull used for breeding in the region. Yet they reported that there were only "a few pure-bred animals in Mexico." "Most of the livestock however, is of mongrel breeding," they noted: "The beef cattle which roams the range in Mexico are mainly of mixed breeding but usually show some blood of the original Spanish cattle." This interest in mapping the genetic quality of Mexico's cattle demonstrated the commission's conviction that the highest-quality animals were ones that had been uncontaminated by the intermixing of genetic difference.

This kind of focus on the genetic inheritance of farm animals, or their "stock," had been widely promoted by the US Department of Agriculture and land-grant universities in the early twentieth century. As historians of American agriculture have shown, "better breeding" campaigns in animal agriculture shared much with eugenics—the idea that human reproduction should also be managed to produce more desirable traits.[11] Not only did prominent eugenicists look to agricultural breeders to justify their ideas about biological inheritance, but agricultural scientists contributed to the popular uptake of eugenic logics in their efforts to promote "pure-bred" crops and animals. Cattle breeding was one realm in which Americans learned to think about inheritance in a way that normalized ideas about race as a fixed, biological essence that could be handed down through generations.[12] Even as Americans began to turn away from eugenics after it was associated with the atrocities of Nazi Germany, animal breeding continued to be a conduit through which durable ideas about race as "trait" continued to circulate. Along these lines, the commission's comments about the blood purity of Mexico's cattle conveyed a view of animal heredity that paralleled more damaging eugenic logics in which some human groups are racialized as biologically inferior.

As agricultural scientists, the members of the Survey Commission had built their careers in the American institutions in which a cross-fertilization between seeing "race" in plants and animals and racializing humans had proliferated. So, it is unsurprising that they also displayed a keen interest in the heredity of Mexico's people. Several of their photographs of indigenous people suggest that the commission's views of cultural development were linked to ideas about biological inheritance. Two pictures of Indians delineated conflicting paths of development. "Some of the Indians are quick to learn new crafts," but others are "also quick to learn bad habits": "beggars, such as this Otomí Indian, are a common site

Figure 6. Students at the school of Huichapan, Mexico, 1941. Photo credit: Rockefeller Archive Center.

wherever the tourist goes." Alongside photos of young men at an agricultural school they visited in central Mexico, they wrote: "The students at the school at Huichapan are principally Indian boys of the Otomí tribe. Their complexions contrast sharply with the faces of the four members of the commission" (figure 6).

Even though they drew a direct link between Mexico's large number of Indians and what they viewed as its extreme poverty, they held that cultural development of the country's Indians was possible. For example, another photo's caption reads: "The "mestizos," an Indian-Spanish mixture, dominate the Mexico of today. The Governor of Tamaulips, pictured here, was once a peasant. A new hybrid race of people is in the making in Mexico." In explaining the governor's transition away from being "a peasant," the commission put forth a developmental teleology in which the majority of Mexicans were progressing away from a racial identity marked by "Indianness" and toward one closer to Whiteness.

The "new" race they envisioned was, of course, nothing new. It could be traced back to the beginnings of Spanish colonization and the conquistadors' widespread sexual exploitation of Indian women. But the commission's views were consistent with those of prominent Mexican nationalists that constructed mestizaje as a national resource.[13] Modernizers within Mexico viewed this kind of "hybrid race" as a transitional phase in which culturally inferior Indians progressed toward being more cultured mestizos. In their contrasts between Indian poverty and cultured mestizos, the commission adopted this view. From its very beginning, the MAP project would be bound up with the idea that Mexico's poverty was Indian poverty, and that its advancement would be realized through the modernization—defined here as cultural improvement—of its indigenous population.

The question of whether the MAP would realize this kind of development would largely revolve around whether they could develop crop improvement projects that could reach the country's millions of small-scale farmers. Unlike the United States, Mexico had millions of acres under collective ownership. During his term in office (1935–1940) Mexico's president Lazaro Cárdenas distributed between eighteen and twenty million hectares of land to around 750,000 people. Indeed, the Cárdenas years saw 65 percent of the overall distribution of land during the reforms that occurred between 1917 and 1940.[14] Cárdenas instituted agrarian reform and set up collectively owned village lands for food production, or ejidos. During the Cárdenas administration, relations between the United States and Mexico were often strained. Cárdenas's efforts threatened US corporate interests, causing tension between the two nation-states. Some of the land seized as part of his redistributions were estates owned by Americans, and in 1938, he moved to nationalize the oil industry, a policy that included seizing Standard Oil properties, which, coincidently, were the source of the Rockefeller family's wealth.

The Survey Commission members were not socialists and were not strong advocates of collectively owned farmland. At the same time, they seemed inclined to work with the present conditions in Mexico and were interested in developing the ejido system. The commission did not offer extensive comments on the ejidos, reflecting that the system was "still on trial." "It appears to operate successfully in some areas, but there have been numerous abuses and many failures," they concluded. Though the members of the commission were by no means experts in Mexico's political situation, their comments suggest that they found the country's revolutionary tradition too "political." Commenting on Mexico's agricultural schools, they wrote that the need was to "take school out of politics and put science into the school."

Training a New Generation of Scientists

The Survey Commission's recommendations, which were adopted without change by the Rockefeller Foundation Board of Trustees as "the guideline for an action program in Mexico," would make the mission of improving agricultural schools central to its program. However, the committee suggested that "[Mexican] schools can hardly be improved until extension men are improved [and] investigational work cannot be made more productive until investigators acquire greater competence." Though Wallace had called for developing Mexico's corn and beans, the commission concluded that any development project in the country would first need to improve Mexico's agricultural scientists. The MAP would emphasize training Mexican agronomists throughout its two decades. With Rockefeller Foundation funding, over 450 Mexican scientists would conduct scholarships and fellowships at the MAP. Those deemed the most "intelligent [and] industrious" were selected for fellowships in American universities. By 1959, about one hundred scientists had been sent to study at US schools, mostly land-grant universities (LGUs).[15]

In his role as the foundation's most senior agricultural advisor, Stakman would advocate for the MAP's efforts to train Mexican scientists. Reporting to his colleagues on the foundation's board of agricultural consultants, Stakman argued that it was the MAP's training efforts that had truly sparked an agricultural revolution in Mexico. After returning from a visit to the MAP in 1953, he was pleased to report that the program had molded Mexican agronomists into "competent scientists and cultured men."[16] Crediting their exposure to American scientists as the key factor in their advancement, Stakman even suggested that working alongside Americans had guided the Mexican scientists to choose more "cultured wives." This comment about "marrying up" suggests that ideas about cultural improvement were tied up with ideas about biological reproduction (better breeding). In a later assessment of the MAP's education projects, Stakman expressed similar concerns with the racial "stock" of Mexican scientists.[17] He argued that the MAP had brought about widespread change but that progress had not come easy. Working with Mexico's agronomists had been difficult because of their unique "social psychology," which he defined in racial terms. He described "a distinct Mexican race" that "combined the intense pride of the Spaniard and the impenetrable inscrutability of the Indian." Using the language of biological inheritance, Stakman argued that Mexicans suffered from an "inferiority complex" and were more likely to be "introverted, hypersensitive, and uncertain." Though he claimed it was "too late to do much about Mexican genes," he argued that the foundation's work to change the educational environment might do enough to overcome Mexicans' inherited shortcomings.

Stakman and other Rockefeller Foundation leaders credited the MAP's extensive American fellowship program, which sent the best and brightest agronomists from the program for graduate-level training at LGUs in the United States, as critical to launching what they saw as a "revolution" in Mexico's agriculture.[18] Historians have shown how the influence of the LGU system led American scientists to favor a production system that benefited Mexico's commercial growers, while failing to substantially contribute to improving the plight of campesinos. But the relationship between the LGU system and the MAP's work begs further attention—not only because it was incompatible with indigenous knowledge systems, but because it was built upon appropriating them.

In an extensive report, Robert Lee and Tristan Ahtone show how LGUs used massive amounts of land granted to them by the US government to rapidly grow their endowments.[19] The 1862 Morrill Act granted millions of acres of land to each state to use for the purpose of generating money toward endowing their agricultural colleges. This land was in many cases far from the actual sites of colleges, extending far and wide across the lower forty-eight states. Much of the land was in the public domain because it had been taken from indigenous peoples through treaty, land cession, or seizure. In some cases, the people kicked off their land had only recently been dispossessed. American LGUs were then built "not just on

Indigenous land, but with indigenous land." Some of the land was sold. Some of it was speculated upon. And some remains held in trust. All of it has generated revenue that continues to fund the massive endowments of LGUs today.

The LGU system was expanding through expropriated indigenous land while it denied the contributions of indigenous people, land, and cultural material. It was, as Rod Ferguson has argued, a deeply exclusionary project.[20] Blacks and Native Americans were excluded from the 1862 Morrill Act's promises of expanding democratic education to rural America. American scientists working at the MAP were part of the land grants' lineage. In their efforts to create a new generation of Mexican agronomists through the LGUs, they also connected their Mexican collaborators into an internationalizing agricultural knowledge project rooted in the LGU model. Many former Rockefeller Foundation fellows went on to hold important positions in Mexico's Ministry of Agriculture, agricultural universities, and at agribusiness companies. But the MAP did not just "export" American agriculture. It also reproduced the political, cultural, and knowledge systems upon which that agriculture was built. The project reflected American settler colonialism as an ongoing process rather than an event in the past.

Of course, debates over land use in relation to indigenous peoples were much different in Mexico than in the United States. In fact, the MAP came on the heels of massive efforts to expropriate land not from indigenous peoples, but from wealthy landowners, and grant that land back into the hands of Mexico's dispossessed. Yet, though some MAP scientists were sympathetic to the plight of the campesinos, their broader perspective eschewed politics in favor of what they understood to be the more important technical, scientific approaches to agriculture. Along these lines, MAP leadership expressed their preferences for working with younger Mexican agronomists who were less influenced by the history of the Mexican Revolution.[21] Connecting the legacy of the LGUs to the politics of land shows that though they claimed to be apolitical, their science was a deeply political project. In addition to devaluing indigenous knowledge and land, the MAP expanded through a system of appropriating the ancestral homelands of Native peoples, while disavowing the contribution of that land. In this way, the story of the LGUs is one that epitomizes the orientation toward indigenous people and land as "possessive"—one that appropriates land and material culture while, at the same time, disavowing how those "resources" are central to its knowledge production. We get a better sense of how this orientation was produced through the MAP by looking into its expansive project to collect and distribute maize from across Mexico.

IN SEARCH OF MAIZE'S HIDDEN WEALTH

Though accounts of the Green Revolution typically focus on its wheat program, the MAP's efforts to improve Mexico's maize would also generate lasting ramifications. In the program's first years, the Rockefeller Foundation would hire E. J. Wellhausen, a corn geneticist with a PhD from Iowa State University, to lead its maize project.[22] After leading the maize project in the MAP's first decade, he took over as director of the entire program from 1951 to 1959. He later became the first director general of CIMMYT. Wellhausen led efforts to collect and categorize hundreds of varieties of maize from throughout Mexico and Central America. Wellhausen's work not only contributed to the material foundation of a long history of maize breeding across the world's tropical areas, but also had far-ranging consequences for the way the crop would be used. In a 1966 oral history recorded by the Rockefeller Foundation, Wellhausen speaks about spreading his maize collection across a warehouse floor, mapping the country in terms of where each variety was grown. He then bred the maize together and reduced his collection of around six thousand varieties to around one thousand "composite" varieties. He distributed these composites to breeding programs around the world, including development institutions in Kenya, India, and Thailand and US seed companies operating in the American Midwest and throughout the Southern Hemisphere.[23]

After arriving in Mexico, Wellhausen immediately began collecting as many samples of maize as possible from across the country.[24] Understanding Mexico as maize's "center of origin," Wellhausen viewed the country's "exotic" maize as a wealth of "undiscovered" varieties that might prove advantageous for commercial breeding.[25] In his oral history, Wellhausen discusses numerous "collecting trips" he and his colleagues took in search of maize. He recalls speeding across Mexico's highways, stopping only long enough to collect maize from roadside fields, granaries, and hillside farms:

> We collected a lot of corn, right along the road, as we went from one place to another, along the main roads in Mexico. This was how we got the first corn collections we ever made. And this was very interesting to me. This introduced me to the people that were growing the corn and the Indians, and so on (29).

Like the Survey Commission, Wellhausen comes to understand Mexico's agriculture by mapping the country as a geography marked by Indianness—by "locating Indians in the landscape."[26] Though it appears to be an offhand comment, Wellhausen's depiction of Indians as separate from "people... growing the corn" can be read as encapsulating the MAP's epistemological foundation. As the Survey Commission had recognized, many of Mexico's maize farmers were indigenous. But Wellhausen's scientific approach toward the plant isolates it from its cultural context, confining indigenous peoples to the background of his project. He is "introduced" to maize, farmers, and Indians as objects for his own knowledge, rather than actors that directly contribute to that knowledge.

This scientific detachment is further illustrated by the fact that Wellhausen hardly spoke to the people they encountered on collecting trips. Detailing how he employed Mexican college students to facilitate their collecting efforts, Wellhausen recalled: "We stopped along the roadsides and collected corn. The young men I was with did the talking and made the arrangements with the people that we collected corn from, and I learned a lot about corn" (32). This direct language conveys Wellhausen's singular focus on corn. People that grew corn were only useful insofar as they could facilitate his access to it. Wellhausen himself found little interest in venturing into this background. As the collecting mission expanded, he mostly relied on the college students to collect his samples. He would give them bus tickets and some empty sacks and instruct them to go out into the "hinterlands" and collect maize from the people they found there (133). Wellhausen utilized one agriculturalist in particular, Efraín Hernández Xolocotzi.[27] (The Rockefeller scientists did not use Hernández's Aztec name, calling him "X" for short.[28]) Wellhausen praised "X" for his ability to speak "Indian dialects," interact with "the most remote villages" on behalf of the project, and negotiate with Indians to gain access to their maize, even ceremonial varieties (133, 136).

Wellhausen and his colleagues had amassed over two thousand varieties of maize by 1950. Wellhausen was especially interested in varieties that might yield useful traits for breeding efforts and described his work in terms of a hunt for these beneficial genes. Discussing one trait believed to yield more drought-tolerant maize, he remarked: "Not very many people know that this gene exists" (60–61). Wellhausen acknowledges that Indians played a role in the gene's evolution: because they replanted seeds from plants that survived droughts, they had "no doubt" selected for the characteristic. But, situating their farming practices in the past tense, he argues that Indians had "never really fixed [the gene] in the variety" (61). White Earth Ojibwe historian Jean O'Brien shows how American settlers in New England told "firsting, lasting, and replacing" stories about Indians. These narratives erased Indian history and created an enduring myth about settler modernity and the "vanishing Indian" in the process.[29] Exhibiting a "firsting" logic, Wellhausen claims to be one of the first people to recognize a particular gene, claiming a decisive break from Indian practices in the process. Indians are understood to be living relics of the past. Their cultivation of the physiological dynamics of maize plants that Wellhausen names "gene" can only serve as a precursor for his modern science.[30]

This kind of "firsting" logic would underpin Wellhausen's frequent arguments about integrating "exotic" maize into commercial breeding efforts. During a talk he gave at the 1965 American Hybrid Corn Industry Research Conference, he claimed that "the possibilities for the further improvement of corn through a more complete exploitation of the many different germplasm complexes existing in the tropics [was] extremely great."[31] Speaking of the potential of "some of the outstanding indigenous varieties of Mexico," Wellhausen posed a rhetorical question to the agribusiness officials at the conference: "If this is what man and nature have produced in a more or less haphazard way through chance inter-hybridization of different varieties and races, what can the modern geneticist do with his present knowledge of genetics and gene action, and with over 300 different races at his disposal?" Though unnamed, Indians are once again central to Wellhausen's argument. As the absent "non-fixers" of genes, Indians are figured as part of nature. The modern geneticist, by contrast, is uniquely capable of standing apart from nature to see and manipulate genes.

Wellhausen exhibits a "possessive" orientation toward maize that is both material and epistemological, claiming the seeds, but also all the previous efforts of "man and nature." Indigenous scholars argue that settler colonialism extends not only through the logic of eliminating indigenous peoples, but also through a logic of possession.[32] Native Hawaiian historian Maile Arvin defines this logic as "the claiming of other peoples' bodies, identities, and other resources as one's own, without regard to those peoples' own histories and desires for the future."[33] As Arvin shows in a history of how twentieth-century scientists racialized Polynesians as "almost white," the logic functions "through whiteness." Along similar lines, indigenous Australian scholar Aileen Moreton-Robinson conceptualizes a "white possessive . . . mode of rationality" through which White people understand land and indigenous peoples as objects existing in a "state of nature," only to be possessed.[34] The argument Wellhausen made at the seed industry conference is one in which "Indigenous 'others' are represented and constituted in discourse as white epistemological possessions."[35]

Wellhausen's possessive logic is also evident in the book based upon his maize collection: *Races of Maize in Mexico: Their Origin, Characteristics, and Distribution*.[36] This Harvard University Press book described the MAP's maize collection in terms of racial categories. The book's authors wrote that the project presented some difficulty because, as an open-pollinated plant, maize intermixes widely and unselectively. Though they acknowledged the futility of searching for "pure" races of maize, they maintained that "races" could essentially be deduced through categorizing plants based upon their phenotypic characteristics. Following American botanists Edgar Anderson and Hugh C. Cutler, Wellhausen and his co-authors defined "race" as "a group of related individuals with enough characteristics in common to permit their recognition as a group."[37] Using this definition, they charted twenty-five races and four sub-races of Mexican maize. While most of the varieties in the collection were combinations of different "races," they outlined likely "genealogies" of different racial groups in family tree–like diagrams.

The book's language and diagrams suggest a way of racial thinking that extended beyond plants. Indeed, Wellhausen and his colleagues' efforts to map the races of maize were surely entangled with their perspectives on human racial improvement. This does not mean that the two strands of conceptualizing "race" were one and the same. Their work to construct a taxonomy of maize and their ideas about human race were "co-produced."[38] Scientific practices always exist within political, cultural, and social systems. But their relationship to those systems is not a one-way street. Science also produces the social. Even though the authors of *Races of Maize in Mexico* were primarily interested in the plant's physical traits, they portrayed a taxonomy that both paralleled and reinforced contemporary ideas about human "race," especially those concerning Mexico's indigenous peoples. For example, their evolutionary history of maize aligned with the kind of eugenic and racialized ideas about improving the cultural standing of Mexico's indigenous peoples that both Mexican nationalists and American scientists promoted. The book's four general racial subcategories are situated in terms of a history of Spanish colonialism, with maize varieties defined as "Ancient Indigenous, Pre-Columbian Exotic, Prehistoric Mestizos, and Modern Incipient." The categories parallel racial improvement logics, in which indigenous people progress by intermixing with European "blood." (Similar to the Rockefeller Foundation Survey Commission's views on mestizos.) Crucially, both ways of thinking about—and applying—the concept of race functioned to reinforce ideas about indigenous inferiority. They also worked to minimize indigenous peoples' work developing maize traits while foregrounding the agency of the "modern" (non-indigenous) breeder. The key point, then, is not so much that the scientists could not disentangle their understanding of the races of maize from ideas about human race, but that the "co-produced" practices of using "race" discounted indigenous knowledge and naturalized the idea that indigenous cultural heritage could (and should!) become the possession of international development breeders—and soon thereafter—corporate seed companies.

This underlying logic contributed to the MAP's inability to develop a maize project that would benefit Mexico's smallholder farmers. Early in the project the MAP maize breeders were sympathetic to the plight of Mexico's mostly indigenous smallholder farmers, or campesinos. Because of this, they focused their attention on developing improved varieties of maize that were still open-pollinated—ones that farmers could replant in subsequent years—instead of hybrid varieties that they would need to purchase new each season. However, partly influenced by their collaborators in the Mexican ministry of agriculture, who urged the need to develop more hybrids, the MAP eventually turned away from developing open-pollinated varieties. By the mid-1950s, the program was growing almost exclusively hybrid seeds.[39] As the MAP neared the end of its second decade, its corn project would focus on "major production areas first to supply superior varieties that can contribute large amounts to national production."[40] Yet even as the program catered to Mexico's more industrialized growers, the uptake of hybrid seeds at the national level was modest. Researchers have shown that by 1963 (the year the MAP's successor, CIMMYT, was founded) only around 12 percent of Mexico's maize was grown using hybrids.[41] Eventually, both Wellhausen and the members of the Survey Commission would express regret that their work had not done more for Mexico's millions of smallholder farmers.

Despite their later regrets, the American scientists working at the MAP and its institutional off-shoot, CIMMYT, largely worked to develop a model of commercial agriculture that was decidedly not in the interest of campesinos. The first director of the MAP—and later director of the Rockefeller Foundation—George Harrar, would ultimately conclude that one of the best things the organization had done was make Mexico more amenable to private agricultural companies, especially American firms.[42] Their ongoing collaborations with American hybrid seed companies also illustrates this outlook. Rockefeller Foundation records show that MAP scientists hosted leaders from American companies and attended conferences alongside seed company officials. Wellhausen went so far as to mail a copy of Mexican seed laws to representatives at DeKalb, offering them advice on how they might move into a country that had been historically opposed to American corporate expansion. To the president of the Illinois-based seed company, Wellhausen wrote: "I strongly believe that the only way we are going to get any volume of hybrid seed used in Mexico will be with the aid of organizations like the one you represent."[43] Along these lines, Elmer Johnson, a maize geneticist at CIMMYT and director of the organization's Central American Maize Improvement Program, spoke about sharing seeds with American companies Pioneer, Dekalb, and Northrop King.[44] Though these American-based multinationals would not make inroads in Mexico's maize sector until the 1970s, their earliest trait development and knowledge of Mexico's conditions were aided through collaborations with officials like Wellhausen and Johnson.

Similar exchanges of knowledge and seeds also helped a burgeoning American hybrid seed industry extend beyond US borders. For example, Wellhausen described contributing maize seeds for Pioneer's first research program outside the United States, which the company began in Jamaica in 1964. Those seeds helped establish roots that helped the company reach a global scale. Pioneer founded its "overseas" subsidiary in 1970 and quickly spread its operations throughout Central and South America in the following decade.[45] The genetic material adapted from CIMMYT seeds would be shared across the company's growing global network of "experiment stations," where they would develop "lines" adapted to regional climates. Though Pioneer and the other largest American seed companies did not scale up their global production of hybrid maize until the 1970s—after the public sector had "paved the way," as one researcher puts it—they would quickly enlarge their global footprint.[46] By the late 1980s, the major US multinationals (Pioneer, DeKalb, and Cargill) "each test[ed] hybrids in 90 to 100 countries and [had] experiment stations in 15 to 20 countries."[47] The genetic material shared through MAP and CIMMYT transfers were crucial to this process. Thus, even if Wellhausen expressed regret for not doing enough for smallholder farmers, his appropriation of their seeds would feed into the breeding "pipelines" of rapidly expanding American multinationals like Pioneer. Jack Kloppenburg argues that CIMMYT and other CGIAR centers facilitated the transfer of plant genetic material from the South to the North, as Western agricultural scientists claimed plant genetic material as the "common heritage of mankind."[48] But this concept of the world's germplasm as "common heritage" should also be thought of in terms of a possessive logic that discounts indigenous knowledge.[49] Maize breeders working across the world's tropical areas would ultimately benefit from the genetic material extracted through the MAP's expeditions into Mexico's backcountry (figure 7).

Since moving into Mexico in the 1970s, multinational maize seed companies have increased their market share. Following the path that Pioneer and its leading competitors first laid out in the United States, they have grown their presence in Mexico through mergers and acquisitions.

"THE LANDRACES ARE IN THE HYBRIDS"   81

Figure 7. Examination of indigenous varieties of corn, Mexico, 1959. Photo credit: Rockefeller Archive Center.

Today, the largest producers of commercial maize in Mexico are the multinationals Corteva (formerly DuPont Pioneer) and Bayer (formerly Monsanto), followed by three of the biggest Mexican companies. However, most of Mexico's farmers do not purchase maize seeds from these companies. This point was made clear to me when I attended a one-day, "Seed Security for Food Security" symposium sponsored by DuPont Pioneer in 2016. The program was a side event of the World Food Prize conference I described in previous chapters. It was held in Des Moines, in the heart of the American Corn Belt. During the afternoon panel in the conference rooms at the downtown Des Moines Marriott, I listened to speakers from DuPont Pioneer and CIMMYT talk about the need for partnerships between their two organizations. Displaying a large, color-coded map of Mexico, a CIMMYT official reminded the Pioneer representatives in the room that most of Mexico's maize is grown with open-pollinated varieties—only around three million of the eight million hectares of maize were grown with certified hybrids. Using a laser pointer, the CIMMYT official circled large swaths of the country's central and southern regions that were shaded in yellow and pink hues on the map. "This is the kind of opportunity we have in Mexico," he said.

These untapped regions on the map represented not only the frontier for capitalist seed companies like Pioneer, but also a kind of epistemological frontier that marked the boundary between where seeds are viewed as cultural resources and where they are understood to be commodities. While both CIMMYT and Pioneer officials stressed the need to "conserve" the maize Mexico's indigenous people grew, they reiterated a narrative about conversion that Green Revolutionaries have been telling since Borlaug: once smallholder farmers witness the power of seed "technology," they will "convert" to buying hybrid varieties (expanding commercial seed markets, in the process).[50] If indigenous growers are to maintain their seeds, it will be through specialty markets, in which "single maize" tortillas are sold to elite consumers in urban areas (along the lines of "single-barrel" bourbon). Yet this scenario, in which campesinos must either become hybrid maize customers or sell their maize in niche specialty markets ignores a whole range of other political—and ecological—possibilities for Mexico's maize growers. Even as corporate-backed modernization schemes continue to target Mexico's indigenous growers, a rising coalition of social movements and farmer groups are pushing back against the hybrid maize model. Calling for food sovereignty, these groups insist on the right to practice indigenous or campesino agriculture as such.[51] In doing so, they also challenge an enduring script about the power of "modern" seeds and the inevitability of a corporate takeover of seed systems. But, as the officials at the Seed Security symposium made clear, this is a story that transcends national and continental borders. The seeds grown by companies large and small across Africa were developed using the "lines" passed down from the earliest Green Revolution project. Yet the appropriation of indigenous maize would not be a one-off event. Indeed, the continual ability to utilize genetic material from landraces in contemporary development would prove essential for the ongoing Green Revolution.

CONCLUSION: BACK TO THE FUTURE IN THE SEED BANK

Maize is the central crop of today's Green Revolution in Africa. Not only is it the most consumed food crop on the continent, but it is also the most important commercial crop for seed companies—from small and medium-sized companies working in particular regions to the world's largest agricultural biotechnology companies. As I discussed in chapter 2, building the capacity of hybrid maize companies continues to be a primary focus of Western-led development efforts across the continent. Funded by USAID and the Gates Foundation, the core Green Revolution in Africa efforts revolve around maize. Some partner with multinational biotech companies to develop GM maize, while others focus primarily on expanding the reach of hybrid varieties.[52]

Climate change is a primary concern for these projects. As maize growers cope with increasingly unstable rain patterns, higher temperatures, and more frequent and severe droughts, projects like Water Efficient Maize for Africa promise to offer a lifeline to farmers on the brink of climate disaster. At the same time, even the most technologically advanced crop development projects are struggling to adapt to the vagaries of climate change. A case in point is the emergence of crop diseases such as "Maize Lethal Necrosis" (MLN). First discovered in Western Kenya in 2011, the disease soon decimated maize crops across eastern Africa, threatening much of the region's commercial seed production.[53] Beginning in 2013, CIMMYT partnered with Kenya's agricultural research system to start a research center that could quickly screen maize varieties in hopes of discovering resistance to the deadly disease.[54] With funding from the Gates Foundation and the Syngenta Foundation for Sustainable Agriculture, they began a widespread effort to screen the maize grown across the region. National seed companies, multinational corporations, and national research programs brought their varieties to be screened. But none of their seeds proved resistant to the deadly crop disease.[55] Desperate to find a solution to the growing crisis, CIMMYT requested samples of landraces from its seed bank in Mexico. It was only in the landraces that the researchers found genetic material resistant to MLN. CIMMYT began providing those seeds to companies and National Agricultural Research Systems so that they could incorporate them into their own breeding pipelines. Mexican maize landraces became vital to the continuation of hybrid maize development in the Green Revolution for Africa.

The development of cutting-edge genomic editing technologies has only amplified this promise of mining landraces in search of lucrative traits like drought or disease resistance. Tools like CRISPR-Cas9 (often shorthanded as just CRISPR), give breeders the ability to make genetic edits with unprecedented quickness and precision.[56] Yet much like genetically modified crops, both the technologies needed to use CRISPR and different applications of the technology are legally protected as the intellectual property of a handful of research institutions and corporations.[57] To enable CIMMYT scientists to utilize CRISPR, the institution entered into a sublicensing agreement with DuPont Pioneer in 2016 (the ag-biotech company holds key patents on CRISPR technology). The first project pursued under this agreement was to further develop MLN resistance in African hybrid maize.[58]

CIMMYT and DuPont finalized their CRISPR negotiations during CIMMYT's fiftieth-anniversary celebration in Mexico. As hundreds of international scientists and agribusiness officials gathered to commemorate the legacy of a central Green Revolution institution, CIMMYT and DuPont leaders would ink a deal to shape the Revolution's future. To mark the occasion, DuPont Pioneer's vice president of research and development, Neil Gutterson, gave a speech about the history of agricultural scientific breakthroughs.[59] He compared CRISPR to a "search" function in a computer's word processor: scientists can use it to quickly locate genetic sequences of interest in the "text" of a plant's genome and then efficiently "delete, edit, or replace" genes. With CRISPR, he explained, plant breeders can knock out unwanted genes, bolster a gene sequence to change a plant from having, say, drought tolerance to *high* drought tolerance, or replace an unwanted gene (disease susceptibility) with a desired gene (disease resistance). Because of CRISPR's powerful "search" capabilities, geneticists can now efficiently find "native" traits in a particular plant, and then move that trait into an "elite" variety that has been highly adapted to a particular place. He used the example of editing MLN-resistance genes from Mexican maize landraces into hybrids conditioned for Kenyan agriculture. This process would take years of successive breeding with conventional methods. CRISPR makes it feasible in just one or two breeding cycles.

To convey for his audience in Mexico the magnitude of the CRISPR technology, Gutterson displayed a slide showing the history of "modern plant breeding." The image depicted a timeline that began when Henry Wallace, the founder of Pioneer, first developed hybrid maize and continued through successive advances in hybridization and plant biotechnology. It culminated with the discovery of CRISPR, which Gutterson proclaimed as the next "game changer" that promised to reshape the industry. The way Gutterson marks the beginning of his maize breeding history with Wallace's hybrid maize (and the founding of his company) displays a "firsting" logic that rehearses Wellhausen's claim to be the first breeder to "fix" a gene in maize. Where the senior maize breeder organized his collections in piles of cobs on a warehouse floor, CRISPR offers today's geneticists heretofore-unimaginable optics of discovery. Though CRISPR certainly promises remarkable advancements in breeding, important questions about who owns, controls, and benefits from the technology remain. Posing these questions is not meant to dismiss the technology's potential, nor to suggest that landraces should not be mined for traits that might benefit today's agriculture. We should, however, examine how technologies like CRISPR can bolster particular approaches to climate change adaptation at the expense of others.[60] Whether landraces are primarily viewed as living plants integral to the food sovereignty of indigenous caretakers or as genetic resources stored away in seed banks and mined with genomic tools remains a pressing political concern.[61] That CIMMYT and Pioneer partnered to breed MLN-resistant landraces into the ag-biotech company's tightly controlled hybrid seeds suggests that logics of extraction, appropriation, and possession persist along the Green Revolution's genomic frontier.

4 Seeing Like a Seed Company

In the summer of 2015, a Monsanto plant breeder told me about discovering the "holy grail" of plant biotechnology. Beginning in the early 2000s, he explained, the company's geneticists experimented with inserting strands of bacterial DNA into the genomes of maize plants. They hoped to grow a new plant that could survive—even thrive—under drought conditions. He recalled the moment in 2003 when his team realized they had found their elusive drought gene: "I have a picture of me in Kansas, when we had just collected the ears from the plants in the first field trials. There's one bag from the non-transgenic plants and two bags from the transgenic plants. And we were like: 'something's going on here!' We were pretty excited."[1]

That extra bag of transgenic, or genetically modified (GM), maize held the promise of a potentially blockbuster "trait" for the company.[2] But, as the breeder noted, the actual genetic sequence they had used was rather unremarkable. For years, scientists had known that this gene—*cspB*, from soil bacteria, *Bacillus subtilis*—functioned as a "cold shock" mechanism. But breeding the gene into a commercial crop like maize so that it might adapt to drought stress constituted a novel application. It was this newly discovered use of the gene that Monsanto could patent as their intellectual property. It is in this sense—the gene-as-property—that the drought gene is central to this chapter's story. Beginning in 2008, the drought gene would be used to launch an international development effort aiming to bring GM maize to smallholder farmers in Africa. The project, called Water Efficient Maize for Africa (WEMA), would command attention on account of both what it was doing—developing and promoting biotech crops in Africa—and who was involved—some of the most powerful corporate, public, and philanthropic organizations in international agriculture.

Funded by the Bill and Melinda Gates Foundation and the United States Agency for International Development (USAID), WEMA would be at the forefront of a new breed of public-private partnerships that aim to make proprietary biotechnology products available to smallholder farmers and public sector researchers across Africa.[3] Alongside several similar initiatives, WEMA would operate through a Nairobi-based nongovernmental organization, the African Agricultural Technology Foundation (AATF). It would bring together plant scientists from Monsanto and CIMMYT, who would work alongside researchers from National Agricultural Research Centers in five African countries: Kenya, Uganda, Tanzania, Mozambique, and South Africa. Under WEMA's "philanthropic mandate," they would collaborate to deliver hybrid and GM drought-tolerant maize to smallholder farmers in sub-Saharan Africa.[4] Monsanto contributed the drought gene and one of its older insect-resistance biotech traits to the project, but it would waive the royalty fees it typically charges farmers and seed companies for licensing its patented technology.[5]

There are multiple approaches through which people around the world ask the question of if and how something can be owned. Different social processes—or property regimes—inform how property is thought of and governed.[6] Agricultural biotechnology in the United States emerged alongside the rise of a legal-economic system in which genes came to be viewed as commodities governed under a private property regime.[7] But in many countries seeds—much less genes—are not necessarily treated like private property. When WEMA began, South Africa was the only participating country with a regulatory system for testing and commercializing biotech crops. Before the project could deliver maize seeds containing the drought gene to farmers in countries outside South Africa, private property regimes in agricultural biotechnology would need to be built from the ground up. So, WEMA would initiate a range of efforts to "build capacity" for a legal system in which private ownership of plant genetic material is the norm.

Throughout its first decade, WEMA was the subject of polarizing assessments. Some heralded it as a successful public-private partnership tackling the intertwined problems of climate change and hunger. Others saw the project as a Trojan Horse furthering corporate control of seed systems under humanitarian guise. Rather than try to prove either of these sides correct, this chapter shows how the project's mission to improve the plight of smallholder farmers and its goal of installing private property regimes mutually reinforce each other. Extending my previous discussion of the Gates Foundation's philanthrocapitalism, the chapter describes how that project extends a longer lineage of an "improvement" logic that yokes progress with the growth of private property.

I interviewed fifty officials across WEMA's institutional network, mostly during the project's seventh and eighth years.[8] Soon after, news emerged that Monsanto had agreed to a buyout offer from the German multinational, Bayer. After a two-year effort to appease anti-monopoly regulators around the world, the merger was finalized in 2018. The new company would be the largest seed and agricultural biotechnology company on the globe. To no one's surprise, Bayer dropped Monsanto's notorious brand name. WEMA has continued under the new umbrella, with a new round of funding and a re-brand as "TELA" in 2018. To avoid confusion, I refer to the company as Monsanto in my discussion of the project up until the merger. That history would begin soon after that first promising field trial in Kansas, when Monsanto's CEO declared a mission to take the drought gene to Africa.

IMPROVING YIELDS, IMPROVING LIVES

Before an agricultural biotechnology company like Monsanto releases a new GM seed product, they spend years refining the technology in their research and development "pipeline." As the drought gene—which Monsanto branded DroughtGard in the United States—advanced through its pipeline, company officials began to consider whether the trait might prove useful far beyond the American Corn Belt. Several of the Monsanto officials I interviewed recalled hearing the company's CEO, Hugh Grant, argue that Monsanto had an ethical obligation to get its drought technology into the hands of African farmers as soon as possible. Drought, he would say, was a "matter of life and death" for farmers in Africa.[9] Other leaders across the company echoed this moral appeal. As one Monsanto scientist recalled, there was a widely felt belief within the company that bringing their drought technology to smallholder farmers in sub-Saharan Africa was simply "the right thing to do."[10]

I would hear more of this sentiment as I spoke to people across the WEMA project. Describing smallholder farmers in Africa as "vulnerable," "resource poor" or "the poorest of the poor," WEMA officials spoke about their work with missionary-like zeal. "WEMA is something that you do because you're passionate about the mission," said a Monsanto official.[11] Because of a particularly fervent enthusiasm for the cause, project participants even call one of their colleagues "the preacher."[12] This dedication was evident in my informants' use of the word *improve*, which they used to depict both the work of breeding better maize varieties and that of bettering farmers' lives. One Monsanto representative stressed that the project would only be successful if it "actually helps a farmer change, helps them improve their life."[13] A Gates Foundation official argued that raising crop yields was the key component to "improving the condition of smallholder farmers."[14] Similarly, an AATF official declared: "For us it's not about the money. We're about the livelihoods. Improvement. That is our focus."[15] And a CIMMYT scientist pointed out that their mission was to "improve the income and livelihoods of smallholders."[16]

This improvement story resembles the narratives that drove WEMA's institutional predecessors in the Green Revolution. As shown in previous chapters, American-led programs across Asia and Latin America attempted to thwart the spread of communism by seeding capitalist agriculture and modernizing the peasantry along the Cold War's frontier.[17] There are important differences between Green Revolutions past and present. Today's projects in Africa, for example, are more focused on uplifting smallholder farmers than those of Norman Borlaug and his contemporaries. At the same time, they share a similar orientation in how they portray the people and places they hope to improve. Rockefeller Foundation officials would argue that the agricultural scientists they sent to Mexico in the 1940s found "a land of poverty and hunger in the midst of substantial undeveloped resources." But with intervention from the best of American science, the country was also "ripe for a surge of progress."[18] A similar logic would underpin subsequent international programs, as the Rockefeller and Ford Foundations and USAID expanded their efforts beyond Mexico.

The architects of the Green Revolution would characterize the seeds, farmers, and farmlands they hoped to develop in terms of holding the potential to become modern through agricultural science.[19] American officials at the International Rice Research Institute in the Philippines in the 1960s used an educational film to narrate their ideas about agricultural development to Filipino farmers. The film contrasted the "modern" international research institute with "traditional" Filipino rice farming, which it described as "primitive, . . . inefficient [and] wasteful of human energy."[20] As Rockefeller Foundation projects spread across Latin America, American scientists viewed local farmers as "deficient" and "lacking," but were also optimistic that their scientific interventions could bring about substantial progress.[21] Through creating sharp binaries of modern/traditional, efficient/wasteful, and developed/undeveloped, these narratives shored up the "long Green Revolution" in its efforts to entrench American security interests and capitalist agriculture.[22] They also drew lines between those who would do the work of "development" and the people and places that were "not yet" developed.

But the roots of the WEMA project's mission to improve yields and improve lives run much deeper than the history of the Green Revolution. Indeed, an improvement logic that ties progress to the development of private property runs throughout the history of capitalism. Historians detail how a doctrine of improvement drove the enclosure of English common lands between the sixteenth and eighteenth centuries. The process of dispossessing peasants from commonly held land gained traction through writings of seventeenth-century British scientists and liberal philosophers, most prominently John Locke, whose famous chapter "On Property" is the clearest articulation of the improvement doctrine.[23] Locke argued that in order for one to possess property in land, one must first "improve" that land and make it productive. These ideas would underpin subsequent capitalist enclosure around the world, as narratives about unused or "wasted" lands justified colonial seizures from India to the British colonies in the New World. In European merchants' rapacious thirst for the land, minerals, and agricultural resources of the Americas, they would connect the expropriation of thousands of European peasants to the expropriation of indigenous peoples across Africa and the Americas from their ancestral homelands.[24]

This ideology of improvement produced lasting ideas not only about the proper use of land, but also about what kind of people could rightfully cultivate that land. As Brenna Bhandar shows, settler colonialism is a process through which concepts of race and property are produced together. "The ideology of improvement is one that binds together land and its populations; land that was not cultivated for the purposes of contributing to a burgeoning agrarian capitalist economy by industrious laborers was, from the early seventeenth century onward, deemed to be waste." Those who held different relationships to land, such as common law or use rights, were deemed to be "in need of improvement through assimilation into a civilized (read English) population and ways of living."[25] Race came to be viewed as something tied to particular ways of relating to land and/as property, causing ideas about race and ideas about property to develop hand in hand. As Bhandar summarizes: "Property ownership was not just contingent on race and notions of white supremacy; race too, in the settler colonial context, was and remains subtended by property logics that cast certain groups of people, ways of living, producing, and relating to land as having value worthy of legal protection and force."[26] The way race and property have been put to work together has far-reaching implications that extend into the present. This is not to suggest that the logic of improvement is transhistorical—that it is a singular thing that looks the same each time and place it emerges. Yet, at the same time, ideas about improving land have been used at different times during the past five hundred years to both expropriate people from their land and also exclude other forms of relating to land in favor of private property.

Necessarily condensed, the preceding account of how an improvement logic propelled enclosure and expropriation of land—and the removal of its previous inhabitants—serves as a conceptual backstory for today's calls for improving the lives of Africa's smallholder farmers with Western biotechnologies.[27] My point is not to say that WEMA operates through the same logic as these earlier projects of colonial or developmental improvement. Rather, I suggest that it rearticulates the improvement logic's core assumptions about what progress should look like.[28] Drawing out this connection prompts us to ask why particular farmers—and their ways of relating to land and seeds—are deemed to be in need of improvement. In calling for a development model organized around private property regimes in seeds, projects like WEMA build upon a much longer history of ideas that link improvement and private property. Of course, the story of expropriating in the name of improvement was never a one-time event. It remains "a centuries-long process that continues into the present."[29] We might, then, consider how narratives of improvement are reworked in and through today's Green Revolution.

TAKING THE DROUGHT GENE TO AFRICA

Monsanto's dream of delivering its drought-tolerant GM crops to smallholder farmers in Africa would not be easily realized. Outside of South Africa, African countries' regulatory systems for biotech crops either did not exist or were in an early stage of development.[30] Even if a particular country were to begin approving biotech crops, the road to regulatory approval—"deregulation" in industry parlance—would take years of effort and large sums of money. To shepherd a new GM crop to market entails not only extensive field trials and market research, but also public relations efforts, training of regulators, and coordinated efforts from scientists and lawyers to prepare and submit regulatory documents to governments. There were also ecological challenges to developing the drought technology in sub-Saharan Africa. The tropical maize varieties farmers planted across the region were different from the temperate maize in which the company had been field-testing the drought gene in the United States. It would take Monsanto scientists several years to develop a market-ready variety of GM tropical maize. Because of the considerable resources it would take to clear these political, financial, and ecological hurdles, Monsanto leadership considered the possibility of partnering with a noncommercial, public sector organization in their effort to bring their drought product to Africa. Such a public-private partnership could not only bring in additional funding but might also generate good public relations for a company that activists had spent years maligning as a greedy agribusiness hell-bent on monopolizing the world's seeds.[31]

Monsanto had previously shared its proprietary technology with public sector researchers for humanitarian projects. But a project around drought-tolerant maize presented particular complications. Previous partnerships developed crops like cowpeas, sweet potatoes, or eggplants. Maize, by contrast, is Monsanto's key commercial crop, driving about a quarter of the company's revenue.[32] The drought technology was also still relatively unproven when the conversation about a project in Africa began. Monsanto would only begin limited commercial releases of their Drought-Gard corn in the United States in 2012. Initiating a partnership around a technology still in the "pipeline" necessitated high-level discussions across the company. As one official recalled, these conversations revolved around one question: "How could we structure a public-private partnership that would do two things: protect the company's business and profitability in the developed world, and yet, share it and make it available in the developing world?"[33]

To pursue this question, Monsanto spearheaded a series of meetings with officials from prominent public sector and nonprofit agricultural development organizations. They led to a planning meeting funded by the Rockefeller Foundation and USAID in Arlington, Virginia, in May of 2005. Dubbed the "Drought Tolerant Crop Initiative," the meeting gathered representatives of some of the leading public sector organizations with several biotechnology company representatives, notably Monsanto and its largest competitor, DuPont Pioneer. Though several crops and geographical areas were on the meeting's agenda, participants directed most of their attention to the possibility of drought-tolerant maize in Africa. The biotech companies had made most of their progress in maize and anticipated that it would be the first commercial crop marketed as "drought tolerant."[34] Public sector organizations like CIMMYT had also been developing conventional (non-GM) varieties of drought-tolerant maize.

Though no farmers or farmer-group representatives attended the Arlington meeting, the figure of the smallholder farmer loomed large. The meeting's declared goal was to explore different models for a public-private partnership so that private sector discoveries might benefit "food-insecure" farmers.[35] Yet just how this goal would be accomplished was up for debate. The meeting summary document reveals concerns about "explicit and underlying tensions" among meeting participants, specifically around the issue of how corporations like Monsanto and DuPont Pioneer might serve the interests of Africa's smallholder farmers. Public sector representatives questioned the extent to which the multinational companies would work to develop small and medium-sized seed companies, which were key patterns for international research institutions like CIMMYT. They also argued that public investments in a partnership that benefitted Monsanto or Pioneer in the name of smallholder farmers would need to find ways to "be accountable" to the interests of those farmers.[36]

On the other side of the table, Monsanto and Pioneer officials expressed reservations about the extent to which these smallholders merited their companies' attention. Because most of them saved seed from year to year and had little reason to buy hybrid seed, they had generally been outside the purview of multinational corporations. At the time of the Arlington meeting, their market penetration across Africa was minuscule: less than 10 percent of seed sold on the continent came from the multinational seed companies.[37] Though Africa represented a kind of "last frontier" for GM crops, with the exception of South Africa, the big companies were not devoting many resources to expansion on the continent.

Despite this lack of attention, Monsanto and Pioneer officials argued that public-private partnerships might catalyze private sector investment aimed at Africa's "poor" and "rural" farmers.[38] Because the continent's smallholders were so far outside commercial seed markets, the companies would not be able to reach them without the public sector's assistance. A partnership might push farmers toward buying hybrid seeds and other commercial inputs. Along these lines, a pressing question emerged from the Arlington meeting: "What public investments can make it profitable for the private seed industry to improve the livelihoods of the poor?"[39] This query conveys a line of thinking that would soon drive public-private partnerships like WEMA. For the private sector to benefit the lives of poor farmers, that improvement work must be profitable. Yet to enable this profitable improvement, public investments must first develop smallholder farmers' capacity as consumers of hybrid seeds. No longer simply outside the perspective of the agribusiness companies, smallholders are figured in terms of their potential to become a profitable market. Yet both the onus for developing those farmers and the opportunity to capitalize on corporate donations that will spur that development lie in the hands of the public sector.

As meeting participants on both sides of the table described smallholder farmers in terms that demanded improvement, ethical claims to improve the livelihoods of these farmers became wrapped up with the commercial aims of the companies. Foregrounding the arguments that would soon be central to the WEMA project, the meeting's summary document notes that "private sector plans for maize development and introduction to profitable commercial markets in Africa may be accelerated by the potential to apply drought tolerance discoveries for humanitarian benefit."[40] Beginning with some of the earliest conversations leading up to WEMA, profit and humanitarian aims were framed as mutually beneficial. The Arlington meeting did not directly lead to a formal project. But, in outlining a potential partnership around the twinned logic of humanitarian improvement and commercial expansion, it outlined a path that would soon be followed.

The next year, officials from the Nairobi-based AATF would visit Monsanto in St. Louis.[41] Their organization, which had recently launched in 2004, was developed to make proprietary biotechnology products available to smallholder farmers. Drought-tolerant GM crops were high on their list of projects to pursue. Monsanto had participated in the three-year process of forming the AATF, but wanted to gauge the new organization's technical and legal capacity for managing a large-scale public-private partnership for drought-tolerant maize.[42] Because they were proposing a project to introduce biotech crops into countries in which their use was not approved (or had not been deregulated), the AATF and Monsanto needed to court high-level government officials. (One former AATF official told me about joining the vice president of Monsanto in a meeting with the presidents of Malawi and Tanzania during the UN general assembly in New York.)[43] Representatives from the AATF traveled across sub-Saharan Africa, pitching the project and formalizing agreements with partner governments.

Around this same time, Monsanto was cultivating a relationship with the Gates Foundation. Though it would soon become a major player in international agricultural development, the Gates Foundation did not even have an agriculture program at the time of the Arlington meeting.[44] However, perhaps by somewhat fortuitous events, the Foundation had recently decided that they were interested in expanding into agriculture and had sought advice on how to do so from Monsanto leadership.[45] Gates Foundation officials visited Monsanto headquarters in St. Louis in 2005 and soon thereafter hired one of Monsanto's pioneering plant biologists, Rob Horsch, to lead its agricultural development program.[46] Encouraged by Horsch, the Foundation formally requested that Monsanto and the AATF submit a funding proposal for a biotech maize partnership.[47] In the Gates Foundation, Monsanto found the perfect donor. As one commentator noted, the Foundation was "less political, less bureaucratic, and more corporate friendly" than donors that had passed on a partnership opportunity at the Arlington meeting.[48] As discussed in chapter 2, the Gates Foundation is the quintessential philanthrocapitalist organization in terms of both how it manages grant making as a business and its overall aims to use philanthropy to open markets for corporations. It would be the ideal backer for a project that coupled humanitarian intervention with the expansion of private property regimes in agricultural biotechnology.

CAPACITY BUILDING FOR THE DROUGHT GENE

The WEMA project officially began in the spring of 2008. It would be one of the Gates Foundation's first agriculture projects. Monsanto's testing of their drought trait in South Africa entered its second year but was brought under the WEMA umbrella.[49] With Gates Foundation support, the project partners had ambitious goals. Yet bringing the project's different institutions together would present challenges. Early on, the project had to reconcile what a Monsanto representative referred to as "cultural differences" between representatives from the different National Agricultural Research Systems (Kenya, Tanzania, Uganda, Mozambique, and South Africa).[50] Yet the biggest obstacles revolved around institutional cultures. While each of the partners had signed on to the project's mission to deliver drought-tolerant seeds to smallholder farmers, there were significant differences between the institutions' "cultures, visions, and business models."[51]

"At the initial stage, we had problems," one AATF official remembered. In particular, differences between public (CIMMYT) and private sector (Monsanto) cultures generated "friction" early on.⁵² As one of the oldest centers in the multilateral international network of public sector research institutions, CIMMYT has provided nonexclusive access to the products of its research and development for over fifty years.⁵³ Monsanto's business model, by contrast, is built around patent-protected ownership of its biotechnology traits and tightly guarded trade secrets about its breeding program.⁵⁴ Each organization's work involves drastically different practices and, as another interviewee put it, cultural "attitudes" around intellectual property rights.

It took some work to unify these different approaches. During the project's first year, the Gates Foundation brought in an independent consultant to facilitate a series of "team building" trainings designed to nudge project members toward figuring out how "[they] were going to do business together."⁵⁵ WEMA officials participated in activities meant to help them unite into a collective project "culture," including singing a jingle about doing things the "WEMA way."⁵⁶ Through this work, project participants were able to weather the early "storming" period.⁵⁷ One WEMA representative recalled that it was ultimately through collectively focusing upon the importance of their work that they were able to come together: "We kept reminding people why we were even around the table. The key point was: don't lose sight of who you want this project to serve. Small-scale farmers. If you keep that as your 'north guiding star,' all other issues can be dealt with."⁵⁸ The imperative to improve the lives of smallholder farmers functioned as a kind of binding creed under which the partners could work together. Several of my interviewees compared this unifying process to a marriage. They spoke of how the partners needed to learn how to live with each other and how, committed to their vows of serving the smallholder farmer, they had merged.⁵⁹ But marriages are also legal agreements. Thus marriage is perhaps an apt way to understand WEMA. The project would not only be about getting different institutions to "trust" each other. Because it revolved around the concept of the drought gene as property, a legal apparatus to manage that entity would be essential. Lawyers needed to hash out how the whole thing would work, in other words.

An Intellectual Property Clearing House

A project like WEMA cannot operate without an institution like the AATF. To understand why, we need to consider how that institution was founded around strict gene-as-property dictates. Explaining the negotiations between the Rockefeller Foundation and representatives from the ag-biotech corporations that led to the organization's founding, Rachel Schurman shows that maintaining control over their intellectual property was of paramount importance for the multinationals.[60] Though they were interested in making their proprietary material available for humanitarian uses, the companies were adamant that they would not be "giving" anything away. As one meeting participant recounted, a biotech company official vehemently expressed frustration with a Rockefeller Foundation official who kept talking about the corporations "donating" their intellectual property. "We aren't donating anything," the biotech representative insisted. The companies, rather, were willing to *license* their material with strict regulations about how and where it could be used. By insisting that they would license their material on their terms, the companies ensured that the AATF would be set up as they saw fit. As Schurman argues, "The [AATF's] structure, operating rules, and codes of conduct [all] reflected the agricultural biotech companies' main concerns."[61] In this vein, the institution facilitates licensing agreements that allow ag-biotech companies to maintain tight control over their proprietary genes. As the organization's flagship public-private partnership, WEMA would exhibit this fundamental dynamic.

Some journalists have said that Monsanto "donates" its biotech traits to the project. But members of the project's legal team clarified that this description is inaccurate.[62] To echo the agitated biotech official, Monsanto is not donating anything. Rather, the AATF licenses Monsanto's biotech traits from the company, and then funnels them to other institutions—CIMMYT, seed companies, and the National Agricultural Research Systems—through sublicensing agreements. The AATF works as the "technology broker," channeling Monsanto's property through the project's network, which is made up of both public and private sector institutions. Monsanto agreed to waive the technology fee for WEMA's two biotech traits (the drought gene, and, beginning in 2013, the company's insect-resistance or *Bt* technology). But any users of these traits must sign licensing agreements that detail how and where they can use them.[63] This model of licensing and cross-licensing material is quite common for multinational seed companies like Monsanto. A Monsanto representative explained that over two hundred seed companies in the United States had licensed Monsanto technologies like their top-selling herbicide resistance trait, Roundup Ready.[64] The company even licenses material to and from its largest competitors. But this kind of legal system in agriculture either does not exist or is just being developed in the countries participating in WEMA. Putting the project's licensing arrangements into practice would, therefore, require new legal channels through which the drought gene could move.

Property relations are always social relations. Constructing the legal architecture necessary to govern seeds as private property, therefore, entails shifting how groups of people using those seeds relate not only to the seed, but to each other. By design, the WEMA project would need to change social practices for scientists, seed companies, government regulators, and eventually the broader public. This is what the project calls "capacity building." Early in the partnership, Monsanto brought in legal experts to train WEMA participants in "confidentiality" issues.[65] Through these trainings, WEMA officials learned protocols for sharing information within the partnership.[66] The AATF also helped the National Agricultural Research Centers develop their own intellectual property policies.[67] The project is also building the institutional and legal capacities of seed companies, conducting trainings on how they can license and then work with biotech traits in their own products.[68]

As the project expanded, more and more people would need to learn to work with the drought gene. Regulators would need to be trained to govern the testing and commercialization of new biotech crops. Seed companies licensing WEMA material would need to develop the capacity to "steward" the biotech seeds. And farmers would have to learn the requirements of buying and growing GM crops. Because the "event" of the drought gene is regarded as private property, even when public organizations use the trait, it remains under private ownership. As one of the lawyers I interviewed put it, as long as "someone else's event" is in a particular seed, that seed cannot be shared as a public resource.[69] The private property regime follows the seed. This is how WEMA's humanitarian mission of improving the lives of smallholder farmers articulates with the expansion of private property. But before WEMA seeds can be licensed, sold, bought, and grown, the legal structures for regulating biotech crops must first exist. Knowing this, WEMA explicitly works to influence regulatory reforms, which will facilitate broader shifts in regimes of ownership, moving control of seeds and seed traits "upstream" to private seed companies.

Pushing the Regulatory Envelope

When the project began, South Africa was the only participating country with a functioning regulatory system for testing and commercializing biotech crops. The project worked to advance regulatory systems for GM crops in Kenya, Mozambique, Tanzania, and Uganda. Importantly, each of these countries had different systems in place—some had more capacity to regulate biotech crops than others. But, from the perspective of WEMA's leadership, each of the countries' regulatory systems needed to "catch up" to those in other parts of the world (where GM crops are deregulated). An early "concept note" articulates this point:

> Unfortunately, the vast majority of farmers in [sub-Saharan Africa] have not even had the opportunity to witness field trials of biotechnology products. This "technology gap" is largely due to a lack of science-based regulatory frameworks that would allow testing and evaluation of new agricultural products and reliable delivery systems to reach resource-poor farmers. It means that the most vulnerable African farmers fall further and further behind their counterparts in the developed world. Unless efforts are made now to begin establishing functional regulatory capacity and equipping seed delivery systems, it is unlikely that farmers in [sub-Saharan Africa] will be given the choice to benefit from drought tolerant technology without an additional decade or more of sequential efforts after its launch elsewhere in the world.[70]

Rehearsing the kind of "yield gap" discourse we examined in chapter 2, the document suggests that both African farmers and governments have "not yet" been developed. This then leads to an improvement mission that is both technical and political. Not just about improving maize, the project must "sensitize African policymakers and stakeholders to the importance of biotechnological improvements in maize and . . . improve regulatory policy." The moral urgency of saving "the most vulnerable African farmers" from "fall[ing] further and further behind" amplifies calls to reform regulatory systems. The AATF exists at the confluence of this twofold improvement mission, holding up African smallholders as the target for its projects while explicitly advocating for changing regulatory policies. Though WEMA policy documents state that the project will adhere to the regulatory policy of each country, its wider goals entail rewriting policies. Indeed, WEMA's mission of "ensur[ing] that the developed drought tolerant maize products will be accessible to smallholder African farmers" cannot happen without regulatory reforms. The route to improving the plight of smallholder farmers, again, becomes synonymous with the project of expanding private property regimes.

The idea that WEMA could be a powerful lever for moving governments came up throughout my conversations with project officials. Several of my interviewees talked about how, in the case of Kenya, WEMA had worked to "push the envelope" of regulatory systems by forcing the state to consider an application to deregulate (approve) a maize variety containing Monsanto's *Bt* insect-resistance trait. Since Monsanto offered its regulatory expertise to the drafting and submission process, the application was also a thorough one. WEMA's application for the deregulation of *Bt* maize was written, in other words, with the legal and scientific expertise of a company with arguably more experience engaging agricultural biotechnology regulators than any other. This meant that the project's *Bt* maize application was well positioned to advance the cause of getting Kenya to deregulate its first GM crops. In an exchange worth quoting at length, an AATF official uses the Kenya case to explain to me the potential for WEMA to "shake up" regulatory systems.

> WEMA works the system. What WEMA has done—and it is a very big contribution—is working the policy and certification and regulatory systems with an *actual* product. That has really changed the ballgame. There is no just saying. It's doing. It has changed a lot. [Kenya Plant Health Inspectorate Service] has changed a lot. And the biosafety authorities in Kenya, for example, [have also changed]. And WEMA has pushed the law also because now when you go talk to the politicians there is a product. [Imagining talking to a politician] "This can help farmers? How does it perform? Come. We'll take you. It performs like this. Really? Okay. Where will we get the farmers?

We need the Biosafety law to be passed. And you guys . . . Okay. Just that? Okay. I'll have the chairman of the parliamentary committee in agriculture come and sit in this meeting." And so on. So it works the system.[71]

In this official's simulated conversation with a politician, WEMA works by pushing the envelope. WEMA can "work" the system is several ways. It can force regulators to consider an actual product (as they had in Kenya) or push members of parliament to pass a regulatory policy for biotech crops (a biosafety law, as they were doing in Uganda at the time). This approach proved successful in Kenya, where the AATF and Kenya's National Agricultural Research System were able to get WEMA's *Bt* maize approved for a series of national field trials in 2016. The product includes Monsanto's proprietary trait but will be marketed under the WEMA brand. Though Monsanto's regulatory team worked on the application, it was officially submitted on behalf of AATF and Kenya's Agricultural Research Organization. This behind-the-scenes role is advantageous to Monsanto. In media coverage and meetings with government officials, it was the AATF—not the controversial multinational—that was depicted as the "driver" of the regulatory application process.[72] The *Bt* trait was approved for trials under the terms of WEMA's humanitarian objectives. Using a humanitarian project to move a country closer toward adopting GM crops was proving to be an effective strategy.

"Monsanto No More"

As I mentioned at this chapter's opening, the German multinational, Bayer, acquired Monsanto in 2016.[73] The deal captured attention not only because of Monsanto's decidedly poor public relations reputation (long a central target of anti-GMO, environmental, and anti-globalization activists, Monsanto became a metonym for "evil corporation") but also because of the way it would make already concentrated global markets in seeds and pesticides even more so. Concerns over market concentration sparked opposition from farmer groups in Canada and the United States and both companies had to sell off parts of their holdings to appease antitrust regulators before the purchase was completed in 2018. Hoping to distance itself from Monsanto's controversial past, Bayer dropped the infamous brand.

The new company was unable to offload Monsanto's legal troubles stemming from a suite of class-action lawsuits that alleged that its flagship herbicide glyphosate (the active ingredient in Roundup) caused cancer. Monsanto lost several of these cases, making Bayer responsible for massive damages. A court ordered Monsanto (Bayer) to pay over $289 million in damages in 2018. Though the amount was later reduced, Bayer's stock plummeted. According to a *Wall Street Journal* article that dubbed Bayer's move "one of the worst corporate deals," the German company could be on the hook for anywhere between $5 billion and $27 billion dollars for ongoing glyphosate lawsuits.[74] Though Bayer initially suffered from the Monsanto acquisition, the company would soon rebound and tout a strong outlook for the future based upon its now industry-leading arsenal of GM seeds and agrochemicals.[75]

No longer associated with the Monsanto name, WEMA continued under Bayer. Though the emblem on their shirts and caps changed, Monsanto scientists would continue to work on the project. At a 2018 meeting in Nairobi, WEMA leadership reflected on the project's accomplishments during its first decade.[76] Though South Africa was the only country to have released a GM trait when it commercialized WEMA's *Bt* TELA maize in 2016, project officials were optimistic that the other countries would soon deregulate biotech crops. WEMA members noted that field trials of their biotech products had recently started in Mozambique and had been conducted in Uganda, Kenya, Tanzania, and Ethiopia. The representative from USAID, Tracy Powell, praised the project's efforts to develop and release over one hundred varieties of hybrid maize. "We have actually helped smallholder farmers improve their livelihoods," Powell remarked. "We hope that the next phase of the project can make the necessary contribution to deliver even more compelling products to the farmers."[77] With this goal in mind, USAID and the Gates Foundation authorized a new round of funding for the project's third phase—now called the TELA maize project. Under the new banner, its main goal would be to move biotech crops toward deregulation and commercialization.[78]

Several years into this third phase of the public-private partnership, the initiative scored its first major biotech victory outside of South Africa. In fall of 2021, Nigeria, which had only joined the TELA project in 2018, announced that it had deregulated the project's "stacked" *Bt* and drought-tolerant variety, becoming the only African country beside South Africa to deregulate biotech maize. An AATF official who also worked as part of Nigeria's National Biotechnology Development Agency noted that this was the first "stacked trait" released in the country "as [Nigeria] takes the lead in Africa towards mitigating food security challenges." While the other countries participating in TELA are at different places in terms of the regulatory process, the project officials remain optimistic that they would soon celebrate even more wins. At the time of this writing, it is difficult to say which country might be the next to deregulate a biotech crop. What is clear is that the WEMA project has substantially built the capacity for regulators, seed industry officials, and public sector scientists to work with biotech crops.

CONCLUSION: (STILL) THE FINAL FRONTIER FOR BIOTECH CROPS

Proponents of expanding agricultural biotechnology throughout Africa have long predicted that legal green lights for GM commercialization are just around the corner. But modifying regulatory systems has proven difficult. Across the WEMA/TELA partner countries and beyond, the battle over the future of biotech crops in Africa is still being fought. It would be premature to declare the mission to take the drought gene to Africa a success or failure. It is possible, however, to outline several likely long-term implications stemming from WEMA/TELA.

Bayer is poised to expand upon the Monsanto business model of developing markets by licensing its biotech traits to smaller seed companies for their own use.[79] This is the model that Monsanto used to market *Bt* cotton in India. In 2015, for example, the company licensed its *Bt* trait to forty-four companies across India. Monsanto did not directly produce cotton seeds, but licensed their GM traits to Indian seed companies, which then bred them into their own varieties. In the years before the Bayer acquisition, Monsanto brought in an experienced company official from India to lead its efforts in sub-Saharan Africa.[80] Along these same lines, an AATF official spoke about how WEMA had strengthened links between Monsanto and seed companies—ties that Monsanto might further develop outside of the WEMA project.[81] And one of my Monsanto informants spoke candidly about how WEMA had helped the company gain useful business knowledge about smallholder farmers that might benefit their long-term commercial interests.[82] Since its takeover, Bayer seems keen on seizing upon Monsanto's early market development work. Notably, the company appointed Monsanto's long-time director of WEMA, Mark Edge, as director of both "partnerships" and "seeds and traits business development" for "low- and middle-income countries."

If a WEMA/TELA variety is deregulated in one of the project countries, Bayer will not directly generate profits when another seed company sells a hybrid seed containing the drought gene. The technology fees that farmers pay for the use of the company's trait have, again, been waived. (By contrast, when a farmer in, say, Nebraska purchases a bag of DroughtGard hybrid corn they pay extra to license the GM trait.) At the same time, if a particular country approves one variety of maize, applications for regulatory approval of other biotech varieties will likely soon follow. The biotech companies have other GM traits they can bring to market, such as those designed to make crops resistant to a particular herbicide. (Monsanto's most profitable—and infamous—biotech trait, Roundup Ready, was developed to make crops resistant to the company's top-selling herbicide.) Only the traits shared royalty-free through WEMA/TELA are "off limits" for commercialization. As one Monsanto official put it, once legal frameworks are up and running, nothing prevents the company from applying to introduce additional biotech products.[83] The "capacity building" that WEMA/TELA has accomplished at multiple levels through different national regulatory systems would likely smooth the process of any future regulatory submissions. Again, the outlook in each country is different, but it is certainly plausible that humanitarian, royalty-free biotech crops could pave the regulatory road for an onslaught of commercial traits to come once regulatory systems have developed.

As I have argued throughout this chapter, the gene-as-property is fundamental to WEMA/TELA. The project's legal edifice, scientific practices, and product marketing all revolve around this central feature. Since 2008, the project partners have made far-ranging changes around working with biotech crops. Public sector scientists contributing to the project are changing their communications practices to conform to confidentiality requirements. Seed companies that will license WEMA/TELA varieties are adapting their breeding and paperwork practices. And farmers will have to learn how to properly plant biotech crops—"stewardship," in industry parlance.[84] Each of these demonstrates ways in which the project changes its partner organizations' orientation so that their agricultural and scientific practices are better aligned with the strictures of private property regimes. The project's extensive "capacity building" also suggests that even if it has failed to achieve all its goals for getting biotech crops approved, it has cleared a path for future deregulation and market expansion.

As it expanded, WEMA/TELA would promote a way of thinking about development in which promoting private property regimes in seeds and raising smallholder farmers' productivity were understood to be the same thing. This perspective echoes more long-standing colonial logics that have perpetuated private property regimes through the concept of improvement. The institutions converging through projects like WEMA/TELA—the Gates Foundation, USAID, ag-biotech companies, African governments, and multilateral public sector research centers—bring their own ideas about improving smallholder farmers' livelihoods. At the same time, the narrative coalescing around the WEMA/TELA project serves to naturalize a particular pathway for bettering the lives of farmers. Along these lines, a CIMMYT official told me that if their projects led to improved seeds reaching smallholder farmers, it did not matter whether they also enabled further market penetration of multinational biotechnology companies. Their role was to help make Africa a more favorable environment for investment. They even suggested that five or six regional seed companies would be the first to be bought by the multinationals.[85] These comments about the need to prime Africa for more investment, even if it meant more corporate concentration, recall the arguments the ag-biotech industry made in early discussions about taking drought-tolerant GM crops to the continent: "What public investments can make it profitable for the private seed industry to improve the livelihoods of the poor?"[86] They also parallel the way Gates Foundation officials described their efforts to transform Africa's seed industry (as I discussed in chapter 2). Public-private partnerships like WEMA/TELA, invite a growing number of people to look at smallholder farmers from this vantage—to start seeing like a seed company.

5 Securitizing Smallholder Farmers on the Front Lines of the Climate Crisis

I was at the 2014 World Food Prize conference when I first heard about a novel insurance technology designed to cover millions of farmers on the front lines of the climate crisis. As I would do throughout my research for this book, I had driven to Des Moines to spend a week attending panels and interviewing officials working to bring a Green Revolution to Africa. My research questions centered upon efforts to expand commercial maize markets to smallholder farmers, so a morning panel on "the smallholder's lifeline" caught my eye. As I settled into my seat in the back of the dimly lit Marriott Ballroom, my interest was piqued when Marco Ferroni, the executive director of the Syngenta Foundation, argued that the key to bringing Africa's smallholders into commercial markets was to "bundle" seed sales with a kind of microinsurance called weather index insurance.

Ferroni explained how his foundation—the philanthropic branch of the world's third-largest ag-biotech company—had developed a program in which farmers in rural Kenya used their cell phones to enroll in an index insurance project. When these farmers bought a five-kg bag of maize seed from their local agro-dealer, they would find a small card with a code that they could enter into their phone, linking it to a nearby weather station and activating the insurance policy. If the weather station measured less than a predetermined amount of rain during the growing season, they would receive an automatic payment for the value of the seeds through Kenya's mobile money system, M-Pesa. The Syngenta Foundation had found that insured farmers were more likely to buy hybrid seeds, such as those sold by Syngenta. Ferroni said that index insurance could also prove lucrative for multinational reinsurance companies (insurers who insure insurers), such as SwissRe and MunichRe. These companies held most of the "risk" captured in his foundation's project, which they could securitize—pool together with risk captured from thousands of other microinsurance policies—as a tradable financial instrument. For farmers, seed companies, and insurers, index insurance promised to be a triple win.

The Syngenta Foundation's pilot project led to two new companies, Agriculture and Climate Risk Enterprise (ACRE) and Pula. Based in Nairobi, these companies use digital platforms such as M-Pesa to link millions of smallholder farmers to broader data and risk-transfer markets. At the forefront of agricultural financial technology (agri-fintech), they have garnered industry awards and glowing coverage in Western media.[1] Their goals are ambitious. Pula, for example, aims to use "technology and parametric insurance to insure the previously unbanked, uninsured, untapped market of 1.5 billion smallholders worldwide."[2] ACRE and Pula scale up in two ways: first, they transform large pools of previously uninsured risk into commodities that can be traded by reinsurance companies, and second, they harvest data from farmers' phones, which they can sell to their "upstream" agribusiness partners. Both firms have partnered with multinational seed companies working in the region, including Corteva and Bayer.[3]

These agri-fintech companies market their products as tools to help farmers manage anticipated climate "shocks." This exemplifies the "increasingly pervasive" way in which Green Revolution in Africa projects are framed around a financialized approach to "resilience."[4] That this logic pervades the new Green Revolution has been well established. Yet I suggest we need to pay more attention to the ways that financialized methods of managing "climate risk" link with the US security state's plans for the climate crisis.[5] As I showed in previous chapters, today's Green Revolution builds upon the legacy of Cold War–era American agricultural development efforts across the Global South, which were deeply tied to US foreign policy. In accounts of the Green Revolution in Africa, the role of the Gates

Foundation often overshadows that of the United States. Yet US power remains a dominant force across the continent, shaping both agricultural policies and the broader economic and geopolitical landscape. Analyzing how the US security state aligns with financialized agricultural development projects shows that today's agri-fintech frontier extends broader historical intersections of race, finance, and empire. How do agri-fintech projects and the US security state both position Africa as a space of perennial crisis? And how do these crisis narratives construct "vulnerable" people and places as new sites for extractive investment?

BUILDING RESILIENCE IN A WORLD OF VULNERABILITY

US intelligence assessments describe the intersection of climate change and food security as a critical global issue. During the Obama administration, a series of reports from the National Intelligence Council (NIC), the US intelligence agency responsible for long-term strategic analysis, conveyed the intelligence community's increased interest in the nexus of food security, climate change, and national security. The unclassified version of a 2015 NIC global food security assessment declares that food insecurity in "many countries of strategic importance to the United States" is likely to increase and that the outlook for countries already experiencing food insecurity is likely to worsen.[6] Climate change would only compound these issues. In its 2016 assessment of climate change, the NIC further warns of the impending effects climate change would have on food security, suggesting that climate destabilization would soon lead to widespread social and political unrest. The report concluded that climate change was "likely to pose significant national security challenges for the United States over the next two decades."[7]

A year later, the NIC further outlined the looming threats of climate change and food insecurity in its flagship quadrennial *Global Trends* publication. Titled *The Paradox of Progress*, the report outlines possible future scenarios in a world of ever more frequent shocks—things such as the Arab spring, the financial crisis of 2008–2009, and also large-scale droughts and flooding events.[8] Shortly after its release, the director of the NIC's Strategic Futures Group, Suzanne Fry, presented the report's scenarios to government and industry representatives at the Global Food Security Symposium in Washington, DC.[9] Appearing on a "Food Security Is National Security" panel, Fry told the audience that it was not just countries such as Afghanistan and Somalia that should expect to see climate-triggered instability:

> When we do our [global] risk analysis . . . I'm not kidding you—there's something like two-thirds of the planet that have risk conditions, literally about 120 countries . . . that have risk conditions that make them vulnerable to a shock that could tee off large-scale instability. You compound that with some pretty profound demographic shifts, technological shifts, and we have been living through climate shifts—and [then] the next phase of these climate shifts. So we've got a great deal of vulnerability in the world.

Fry stressed that the United States would not be able to isolate itself from this increasing vulnerability. Discussing American trade interests, she urged that the United States would need to anticipate how climate events might set off shocks that could ripple through the global food system. The best way to manage these inevitable shocks, Fry argued, was to "invest in resilience."

Fry equates resiliency with strengthening international trade and designates resilience as something that can be cultivated in both "economic" and "natural systems." This conjoining of nature and economy illustrates the concept's genealogy. Many people associate resilience thinking with ideas from ecology, especially the idea of complex adaptive systems. But, as Jeremy Walker and Melinda Cooper show, resilience has become a dominant governing logic across multiple fields in large part because it aligns with another set of influential ideas: those of Austrian economist and philosopher Friedrich Hayek. Known as one of the central thinkers behind the rise of neoliberal ideas, Hayek described economic markets as "complex ecological systems."[10] Examining parallels between two key thinkers, one from ecology (C.S. Holling), and the other from neoliberal economics (Hayek), Walker and Cooper show how these two strands of thought "have ended up merging in the contemporary discourse of crisis response through resilience."[11] This integration of concepts about the natural and economic realms normalizes a neoliberal view of nature and society wherein increasing areas of social life are managed through economic reasoning. As we will see throughout this chapter, resilience thinking predominates across financialized development and climate adaptation. Resilience, as Fry's comments suggest, is also a key concept within national security discourse.

Although the possible futures of the NIC's scenario forecasts are grim—financial collapses, atomic bombs, rampant cyberattacks all appear—*The Paradox of Progress* argues that the future will be much brighter for the resilient. Although the future will bring great danger, consequential trends such as climate change will also yield potential for positive outcomes:

> In the emerging global landscape, rife with surprise and discontinuity, the states and organizations most able to exploit such opportunities will be those that are resilient, enabling them to adapt to changing conditions, persevere in the face of unexpected adversity, and take actions to recover quickly. They will invest in infrastructure, knowledge, and relationships that allow them to manage shock—whether economic, environmental, societal, or cyber.[12]

Investments encourage resilience through exploiting opportunity in a world of constant change. Resilient "states and organizations" must adopt a stance of what Walker and Cooper call "permanent adaptability" to navigate this world of ever-increasing risk.[13]

Resilience rose to prominence across different aspects of US national security policy, especially with the post-9/11 establishment of the Department of Homeland Security. This chapter asks how resilience thinking links neoliberal agricultural development projects (i.e., those oriented around a financial rationality) and the contemporary US security state.[14] Importantly, accounting for the scope of the security state means not only considering its more militarized aspects, but also the numerous ways the state's securitizing impulse transcends militarism and incorporates other forms of governance, such as humanitarian or development projects. As Inderpal Grewal writes, the security state "comes to appear—as the state effect theories suggest—as empire not just through military or global policing but also through 'soft power,' exercised transnationally by particular sets of subjects and processes that gain traction because of histories of white racial, masculinized sovereignty."[15] As the US government agency charged with the "development" project, USAID has long been at the forefront of this kind of "soft power." Recently, the agency has made "resilience thinking" a key "organizing concept." The agency's approach to resilience associates humanitarian governance with neoliberal economic development as it seeks to "increase access to financial services" among people deemed to be the most vulnerable to recurrent "shocks."[16] This connection is especially clear in the policy and strategy documents that outline one of the agency's core areas of focus: global food security.

RISK MANAGEMENT FOR PERENNIAL CLIMATE SHOCKS

The NIC's prescription of resilience as the remedy for shock is central to the US Global Food Security Act, which was voted into law in the summer of 2016 with nearly complete bipartisan congressional support. The law pinpoints global food security as a vital national security concern for the United States.[17] As Jamey Essex notes in his history of USAID, since its Cold War–era origins, the US "development" project has always been closely tied to national security interests. At the same time, Essex shows how in the post-9/11 era, as the agency has faced recurring threats of budget cuts, it has increasingly shifted toward an explicit "national security" framework. Alongside the State Department, USAID has "worked to align their core strategies as well as accounting and other internal practices with those of the Department of Defense."[18] Against this background, the Global Food Security bill's framing as "national security" is more than rhetorical flourish. It demonstrates the ways in which the security state is integrating the "3 Ds" of American National Security: development, diplomacy, and defense. The bill's national security framing also proved to be pivotal for getting it passed. At a moment when Democrats and Republicans in Congress rarely agreed on anything, "food security as national security" was something nearly everyone could get behind.[19]

A central policy objective of the law is to "build resilience to food shocks among vulnerable populations and households while reducing reliance upon emergency food assistance." The five-year Global Food Security Strategy (hereafter, strategy) that outlines how the policy will be implemented lists "strengthened resilience amongst people and systems" as one of its three organizing objectives.[20] According to the strategy, building resilience among those it deems most vulnerable is essential as those populations face recurring droughts, floods, and price shocks. Much like the NIC's scenario forecasts, the strategy identifies these "recurrent shocks and stresses as perennial features, not as unanticipated anomalies."[21] It depicts climate change as precipitating a world of never-ending shock—one that demands farmers adopt a "culture of resilience."[22]

The strategy indicates that the road to resilience is paved with financialized approaches to development. It declares that breakthroughs in digital technologies such as mobile money have made it more feasible to bring smallholder farmers into financial markets, allowing them to "both weather shocks and seize economic opportunity."[23] Heralding a market potential of "an estimated US$210 billion in demand for smallholder finance," the strategy references "tailored financial services, products, and systems" as key objectives. Along these lines, it calls for rolling out more financial tools such as crop insurance, credit, and money transfer technologies. Whereas the public sector has led most of the low number of financial development schemes targeting smallholder farmers, the strategy declares that partnering with the private financial sector "will be particularly essential to promoting sustainable development of the agriculture sector."[24] Importantly, the strategy targets not only individual smallholder farmers, but also governments' financial policies—what it dubs "resilience and risk management policy." This is a key element of the strategy and the Global Food Security Act more broadly: countries partnering with US-led development projects must adopt policies that foster an "enabling environment" for private sector investment.[25] Combining the logic of humanitarian governance with neoliberal economic development, American "soft power" works to open markets for international capital. (As Essex and others show, US foreign policy's bend toward neoliberalism has become increasingly agnostic to the question of whether this market development directly benefits a US-based firm.)

The concept of risk management is central to the strategy's financialized approach to resilience. One of its objectives falling under the "resilience" heading is to "improve proactive risk reduction, mitigation, and management."[26] It describes risk in terms of both "potential and realized" and lists "drought, flood, price shocks, pests, and diseases" as examples. And it lists crop insurance technologies such as index insurance as key risk management tools. Risk is a crucial term in the strategy—appearing as something to be avoided, but also something to taken on and transferred.

Like "resilience," "risk" can be a slippery term. Throughout my interviews with officials from development, agribusiness, and philanthropy organizations working in eastern Africa, I was often reminded that smallholder farmers are "risk averse." A truism in development discourse holds that, given their vulnerability, smallholder farmers avoid risk. They are unlikely, then, to spend much on seeds or take loans. As one agricultural economist detailed, "Risk is an impediment to technology investment."[27] Along these lines, one of the Monsanto officials I interviewed about the Water Efficient Maize for Africa project (the subject of chapter 4) talked about how the drought tolerance gained from the biotech trait might make farmers' ability to invest in hybrid seeds "less risky."[28] Index insurance is promoted as a tool to "de-risk" the process of providing loans to smallholders, thereby making it easier for lenders to extend credit to farmers and for farmers to invest in agricultural inputs such as hybrid seeds and fertilizer.[29] Index insurance companies such as ACRE and Pula market their products directly to banks and microfinance institutions as tools that will enable them to extend credit to farmers previously deemed too risky to offer loans.[30]

In the context of frequent droughts and increasingly unpredictable rainfall patterns, "climate risk" is obviously something smallholder farmers want to avoid. Yet much of the discourse about index insurance describes risk as a kind of untapped opportunity. This two-way definition of risk demands that we come to terms with the way risk becomes a commodity. In his history of risk, Jonathan Levy traces the emergence of the modern use of the concept to an international trade in "risks" that emerged in eighteenth-century maritime insurance.[31] He shows how merchants essentially dealt with two kinds of commodities during their voyages across the Atlantic: the first were physical commodities, whether cotton or the human cargo of slaves, and the second were financial commodities, or "risks," that quantified the possibility of losing their physical commodities. Importantly, this second commodity could be separated spatially and temporally from the original cargo that it secured and traded in financial markets. This is the basis for the global trade in risk that continues today. As I mentioned previously, most of the risk in index insurance schemes is transferred to global multinational reinsurance companies such as SwissRe and MunichRe. This risk also holds the potential to be further pooled together as tradable debts—or securitized—in the form of insurance-linked securities.

Understanding how risk functions as a commodity—and as the basis for financial securitization—is crucial to understanding the logic of resiliency as risk management that underpins the Global Food Security Act and index insurance companies. Viewed this way, we can gain a clearer picture of how "investment in resilience" means that farmers "take on more risk"—in the form of both insurance and debt from loans taken out to purchase agricultural inputs such as hybrid seeds. Risk as harm is meant to decrease, for sure. But risk as a relationship between the present and the future that can form the basis for a financial commodity is meant to continuously grow. Tapping into pools of this kind of risk demands farmers adopt new ways of thinking about what might happen each season. Because index insurance is based upon statistical measurements at the weather station or satellite, there is always the possibility of discrepancy between what the index "reads" and what happens in farmers' fields. To transfer the risk associated with drought and crop loss onto an insurance market, farmers must take on the risk that what happens in their field will not correlate with the "trigger point" on the index.

Because of this discrepancy, Leigh Johnson points out, the insurance coverage offered by index insurance is always only partial.[32] Economists call the possibility that farmers will experience drought but still not receive a payout "basis risk." (One economist I interviewed put it in more blunt terms: "It's when the worst thing that could happen to you gets worse."[33]) Because of the issue of basis risk, index insurance creates a twofold dynamic of risk: farmers both offload risk as a commodity to national and international insurance markets and take on the risk that they will face a drought and not get paid. In this process, farmers become not only a particular type of agricultural producer (as they are brought into commodity chains and begin purchasing credit, inputs, and seeds), but also a financial consumer—what Johnson calls "risk-bearing subjects." Thus markets for financial products develop through a process of "expansion by exclusion": insurance coverage increases in a way that demands farmers also bear some of the risk.[34]

The US Food Security Strategy adopts a similar financialized logic, declaring, "Resilience . . . is necessary before individuals can afford the risk inherent in increasing investment in their farms."[35] Index insurance promises to be a tool that offers farmers both "protection" and "promotion"—it gives them ways to transfer risk but also take on more risk as they invest in credit and agricultural inputs.[36] Somewhat paradoxically, this kind of pre-emptive risk management increases risk. This was made clear when I interviewed a Syngenta Foundation official about the index insurance model that developed into ACRE and Pula. "As farmers invest," this official explained, "their risk goes up."[37] From an insurance company's perspective, of course, farmers' "risk"—in the sense of a potential relationship between present and future that can be captured by an insurance contract as a numerical value and a commodity—is meant to always go up. This was the agri-fintech promise that I heard the Syngenta Foundation's Ferroni outline at the World Food Prize in 2014: a lucrative financial frontier bound to expand in the context of climate change.

Because index insurance relies upon indexes to calculate insurance payouts, it should not really be called insurance. More accurately, the financial tool is a *derivative*: "a contract that establishes a claim on an underlying asset—or the cash value of that asset—which must be executed at some definite point in the future. The underlying asset could be a commodity, such as wheat; or another financial asset, such as a bond; or a financial price, for example the value of a currency; or even an entirely non-economic entity like the weather."[38] Because derivatives opened up the possibility for financial speculators to "bet" upon the rise and fall of an increasing array of assets, they have been at the forefront of the financialization of the global economy since the 1990s. Index insurance functions as a weather derivative in that the contract between farmers and insurance companies is essentially a bet on the outcome of uncertain climatic events that may occur in the future.[39] The difference in future values between a scenario in which the farmer receives a payment versus one in which the farmer does not forms the basis for the derivative contract—and for future hedging upon that possible change.

This experimental technology is decidedly aimed at specific geographies. Johnson quotes an industry official who explains, "In developed countries, we don't sell derivatives to individuals. This may be the best we can do in the developing world, but it has implications for consumer protection."[40] This kind of market segmentation, in which more experimental financial tools are deemed only appropriate for the developing world, points to some of the broader ethical questions about the ways in which these technologies might extend legacies of inequalities and violence based upon the marking off of particular people and places as not-yet developed. Given the criticism about rampant financial speculation and unregulated derivatives markets that followed the 2008–2009 global financial crisis and the related food crisis, one might think that there would be some caution on the part of international development organizations about the possibility of "derivatives for development."[41] Yet mainstream development organizations such as the World Bank, USAID, and CGIAR promote index insurance as a socially just means to address poverty and climate change. How could the financialization of the livelihoods of those most vulnerable to climate change be so uncritically promoted—particularly in the wake of the financial/food crisis? Could this be entirely a case of techno-financial utopianism? This is surely part of index insurance's appeal. But it does not fully answer why there is such an exuberance for a financial fix so soon after financial meltdown. Examining how racial geographies shape the way particular places are deemed to be uniquely "at risk" sheds light on how such questionable development practices move forward with little criticism.

RACIAL GEOGRAPHIES OF FINANCIAL EXPERIMENTATION

To think race and geography together entails more than just looking at how a given place comes to be coded as raced—think "inner city" or "Third World" country. It also means thinking about how places take on meaning through social systems structured through race. The concept of "racial geography," then, is not just about locating "a given effect in space in racial terms," but understanding how geographic spaces are made and remade through race.[42] As María Josefina Saldaña-Portillo shows through a discussion of the ways in which the national geographies of the United States and Mexico have been constructed through racialized conceptions of Indians and the land they inhabit, the concept of racial geography points to the historical relations that inform widely held geographical ideas.

Applied to the global geographical divisions that create drastically unequal vulnerabilities to the ravages of the climate crisis, the concept helps us to think through the historical traces behind "*shared* perceptions of space, governed by learned conventions that have developed over more than five hundred years" of colonialism and racial capitalism.[43] It enables us, in other words, to understand how geographical constructions—whether of a "frontier" or an "emerging country"—are produced and normalized. As Saldaña-Portillo shows, history and geography are always intertwined and always play an active role in determining how we relate to a given place. "Geography," Saldaña-Portillo writes, "is not only a discipline for mapping the world to be seen: it is also a way of disciplining what we see, of disciplining us into seeing (and knowing) mapped space as racialized place."[44] This relationship between seeing and knowing "mapped space" in terms of racialized conceptions is especially applicable to Western ways of "knowing" Africa. Indeed, there is a rich archive of Western cultural mappings and representations of Africa—from Joseph Conrad's *Heart of Darkness* (1899) to the pages of *National Geographic* to the web pages of humanitarian organizations—that construct the continent as a unique, separate geographical space.[45] As Kaiama Glover argues, these geographical depictions, so common in journalistic and humanitarian discussions of Africa, work to draw clear lines between "us" and "them." They draw sharp material and ideological borders through which "the 'Afro-' is rendered forever fixed in dystopian time through a disavowal of historical relationships that implicate the West."[46] These kinds of geographical descriptions are commonplace in the discourses of agricultural development and agri-fintech.

A striking example comes from the cover of an influential European-based agricultural development organization's 2015 "Agriculture for Impact" report (featuring a profile of ACRE). Under the headline "The Farms of Change: African smallholders responding to an uncertain climate future," a map of the continent, rendered in heat-map orange hues, suggests that Africa is a dry, cracked, and lifeless parcel of earth. This kind of visualization must be understood as part of more long-standing racial geographies that construct Africa in terms of a dehumanized space. As Glover writes in her analysis of the racist narratives underpinning humanitarian discourses about Haiti and Africa, showing a version of Africa as devoid of life relegates "brown bodies to dehumanized spaces the world over" while suggesting "that survival in such inhuman spaces proves the nonhumanity of their inhabitants."[47] Echoing work of Sylvia Wynter and Katherine McKittrick, Glover points to the ways in which Westerners construct Africa as a dehumanized space of the "racially condemned."[48] For the purposes of my discussion, the key point to consider is the way that these and so many other geographical representations of Africa depict the continent as a site of permanent crisis—where "disaster is a state of being, as opposed to an event."[49] Geographical descriptions of Africa as a singular, separate space persist in the frameworks that underpin the US security state's approach to climate change and food insecurity. The Global Food Security Strategy, for example, references shocks as perpetual events for the continent's smallholder farmers, and the NIC calls Africa a "zone of experimentation."[50]

But what is the relationship between representations of Africa as a unique geography and agri-fintech's appetite for new sources of accumulation? Operating through a frontier logic, companies like ACRE and Pula open sites for profit making along the "risk frontier" of previously uninsured farmers.[51] Their financialization of "climate risk" constitutes a frontier market for the companies and their multinational reinsurance and agricultural input company partners. As Raj Patel and Jason W. Moore write, frontiers are essential to capitalism. "Capitalism," they argue, "not only *has* frontiers; it exists only *through* frontiers, expanding from one place to the next, transforming socialecological relations, producing more and more kinds of goods and services that circulate through an expanding series of exchanges." For Patel and Moore, frontiers are sites at which the "stuff" that generates value for capital—nature, workers, and energy—is put to work "as cheaply as possible."[52] As they show, it is this cheapening of lives, land, and labor that generates profits for capitalists.

This extractive relationship, central to capitalism, depends upon uneven power relations. Race is a key modality through which those relations are produced and exploited. As Jodi Melamed argues, "Capital can only be capital when it is accumulating, and it can only accumulate by producing and moving through relations of severe inequality among human groups—capitalists with the means of production/workers without the means of subsistence, creditors/debtors, conquerors of land made property/the dispossessed and removed."[53] These divisions, "require loss, disposability, and the unequal differentiation of human value, and racism enshrines the inequalities that capitalism requires."[54] Along these lines, scholars examine how this kind of unequal valuation of human lives occurs throughout our contemporary financialized global economy, ranging from the racist mortgage-lending policies that contributed to the 2008–2009 financial crisis to the exploitation of other "subprime," racialized populations in microfinance projects in the Global South.[55] We might also ask how index insurance projects extend through geographical constructions that single out Africa as an exceptional place—a kind of "risk environment" where crisis is endemic. How do agri-fintech projects construct new kinds of "subprime" populations in which to invest while extracting profits largely for the benefit of multinational agribusiness and reinsurance corporations?

In a compelling analysis of how microfinance expands "poverty capital," Ananya Roy explains how microfinance operates through geographical constructions of the frontier.[56] She defines microfinance as "the new subprime frontier of millennial capitalism, where development capital and finance capital merge and collaborate such that new subjects of development are identified, and new territories of investment are opened up and consolidated." Emphasizing the geographical thinking that shapes donor and microfinance institution's conceptions of the subprime subject, Roy makes clear the relationship between an expansive project of poverty capital and the "opening up" of frontier markets. Through this geographical process, the subjects of development of global microfinance are viewed as "subprime borrowers" that are considered high risk: "Their financial inclusion takes place on subprime terms."[57] In the 2008–2009 financial crisis, precipitated in part by a mortgage-backed securities crisis, race proved a critical factor in not only constructing the category of "high-risk" borrowers, but also in allocating blame after the housing crisis. Paula Chakravartty and Denise Ferreira da Silva argue that popular media accounts of the housing crisis as the "subprime crisis" worked to lay the blame on "subprime" racialized populations rather than the bankers who recklessly gambled on their exploitation.[58] They insist that the figure of the "subprime" in both the United States and Global South should be understood "as a racial/postcolonial, moral and economic referent, which resolves past and present modalities and moments of economic expropriation into *natural* attributes of the 'others of Europe.'"[59] Racial logics create particular places and people as naturally inclined to vulnerability and "subprime" economic statutes. This is how the figure at the center of agri-fintech's "financial inclusion" efforts across Africa—the homogenous "African smallholder farmer"—can be thought of as a global subprime.

This subprime logic materializes in index insurance through a range of ways in which development experts seek to train farmers to take on risk. In some instances, the insurance companies find that it makes more sense not to tell the farmers that they are being insured—to simply insure the creditor or agricultural input supplier. This displays the logic of the subprime borrower: farmers do not appreciate the perils of risk, so the financial reasoning should be left to lenders and agribusinesses. This, too, perpetuates racial trajectories in which the "others of Europe" are viewed as subjects "without self-determination."[60] Development organizations likewise deploy a wide range of efforts to both instill a different approach to farming and cultivate a new ethic of resilience in these smallholder farmers. As the Global Food Security Strategy argues, farmers must become more resilient to take on more risk. Financialized approaches to managing climate risk and the US Global Food Security Strategy share this perspective of the subjects of development. They both define smallholder farmers in terms that suggest their unique vulnerability. This construction, moreover, is inseparable from the racial geographies through which agri-fintech and the security state understand—and produce—the crisis.

How might we situate this new racialized subprime figure within the history of finance more broadly? As Zenia Kish and Justin Leroy argue, "Finance has historically developed new innovations through arenas of experimentation in which privatized control over racialized bodies and life possibilities expand the boundaries of financial value."[61] They connect contemporary financial instruments, such as "social impact bonds" and "development impact bonds" which allow third-party investors to make money through funding social programs that serve impoverished communities, to the financialization of slavery in nineteenth-century Britain and the United States. Tracing parallels between social impact bonds and the historical practices that financialized the slave economy, they "examine [these] seemingly unrelated modes of investment to demonstrate that racialized life has repeatedly served as the basis for development of new methods to assess and augment the future value of particular lives."[62] In this way, race has been "a tool with which financial innovators elide the ethical concerns raised by financial practices" across different contexts in the past and the present.[63]

Although index insurance functions as a derivative rather than a bond, its financial experimentation relies upon the kind of revaluation of racialized life that Kish and Leroy examine in social impact bonds. Farmers struggling amidst climate chaos become sites for financial investment, a process that inevitably leads to some farmers being excluded from receiving financial security through insurance.[64] Yet much of the discourse on agri-fintech avoids serious conversation about the ethics of "derivatives for development."[65] I suggest that this reasoning has much to do with the ways in which racial geographies naturalize particular places as "at risk." As one of the economists I interviewed detailed, most index insurance schemes are, perhaps unsurprisingly, set up to minimize the potential loss exposure for the reinsurance companies. This official described how the contract terms for index insurance were largely set up to benefit the multinational corporations calling the shots.[66] So, although the narratives of helping farmers cope with their increasing vulnerability drove the expansion of these projects, in many ways, they are set up in ways that maintain vastly unequal power imbalances between people in the Global North and South. We should consider how these experimental practices cultivate the agri-fintech frontier through racialized understandings of "at-risk" populations and places. To further examine how this framework maintains traction, this chapter's final section turns to an analysis of how the "resilience thinking" of agri-fintech and the Global Food Security Strategy also aligns with the directly militarized aspects of the American security state.

"THE BATTLEFIELD OF TOMORROW, TODAY"

During her talk at the Global Food Security Symposium, the NIC's Fry told the audience that future climate instability meant an increasingly unpredictable global geography of risk. In a world of ever greater "exposure to climate risks," there would be a rise in "dramatic, sudden shock-type climate" events that could bring "catastrophic" changes to global food markets overnight. But even the most sophisticated tools of the US intelligence agencies could not predict where these "climate shocks" might emerge. Climate change precipitated a different kind of risk environment, one that called for, in Fry's words, "building resilience into both natural and economic systems."[67] The always unknowable geographic frontier of climate crisis demands a resilient Security State, one that is adaptable and security focused.

Appearing alongside military officials on the panel, Fry's remarks also conveyed imperial assumptions about the American security state's responsibility to secure a world of climate risk. This logic of fighting an unpredictable threat that might emerge "overnight" anywhere around the globe justifies the expansion of the United States' global military footprint. The NIC similarly considers the threat of climate change and global terrorism. Their forward-looking assessment calls for developing a "resilient" security state. Resilience is imperative because of dispersed, unpredictable threats that make "securing and sustaining outcomes—whether in combatting violent extremism or managing extreme weather" increasingly difficult. Using the "ecosystemic" language of complex adaptive systems thinking, the NIC paints a geopolitical future in which both defense and development are oriented toward a world of constant crisis—rendering a vision of the world in which both the global war on terror and the fight to manage the climate crisis demand preemptive action to secure an ever-expanding frontier.[68]

The NIC's call for developing a more resilient security state and the emerging agri-fintech development paradigm occur during a time in which the US military is enlarging its presence in Africa. The continent has long been considered "off stage" in the American imperial theater. But during the past fifteen years, there has been a steady proliferation of military operations across the continent. Through a series of reports, journalist Nick Turse shows how the US ramped up military engagement in Africa—fighting proxy wars, conducting small-scale counterterrorism missions, training African nations' militaries, and flying drones across the skies over numerous countries—during the Obama administration (2009-2017). This buildup works through a forward-thinking logic, in which US Special Forces recognize Africa as "the battlefield of tomorrow, today."[69] As one

Special Forces official explained in 2013: "[Africa is] exactly where we need to be today and I expect we'll be for some time in the future."[70] This official's predictions have borne out in the time since, as the United States continues to direct more and more military resources toward the continent.

As the US Intelligence Community predicts climate change–caused instability to exacerbate political and social unrest across much of Africa, the United States aims to have a significant, long-term military presence there. Indeed, though it is largely hidden from public debate, Africa has fast become the site of a "sprawling, labyrinthine, and at times chaotic shadow war."[71] Shortly into the Trump administration, in early 2017, Turse published details from internal Pentagon reports that trace an extensive range of secret military bases and "forward operating locations" throughout Africa.[72] Through these bases, US forces can conduct surveillance and counterterrorism operations across the continent. Several African countries also represent strategic hubs for US military efforts in the region and beyond. As African countries emerge as new front lines in the global war on terror, the United States sends more special operations forces to Africa than any other region: "More than 14% of US commandos deployed overseas in 2019 were sent to Africa." As of 2019, American special operations forces were engaged in twenty-two African countries, conducting low-scale counterterrorism combat missions and training forces in partner countries.[73] In a recent report that uncovered previously classified information about the extent of US involvement, Turse notes that the United States has poured billions of dollars into security assistance and established a network of twenty-nine bases that spans the continent. Nevertheless, violence and warfare across the continent has actually increased during the American buildup.

Evoking a new kind of "containment" approach to national security, the Trump administration's official "Africa strategy" identified a rising, "predatory" influence of China and Russia across the continent. Speaking on the occasion of the release of the administration's Africa strategy, US national security advisor John Bolton argued that China's and Russia's expansion in Africa "stunt economic growth in Africa; threaten the financial independence of African nations; inhibit opportunities for U.S. investment; interfere with U.S. military operations; and pose a significant threat to U.S. national security interests."[74] To better dominate what it called the "great power competition" with China and Russia, the Africa strategy called for ramping up US investments across the continent, prioritizing US commercial interests, and increasing military support operations with African governments.

The memorable quote attributed to the leader of US Special Forces in Africa—calling the continent "the battlefield of tomorrow, today"—conveys the kind of geopolitical framework the Trump administration embraced with its Africa strategy. But how does this security logic link up with financial logics of business ventures and climate adaptation practices like ACRE and Pula? Randy Martin suggests that high finance and American warfare in the endless war on terror share a preemptive, securitizing logic, in which "potential threats are actualized as demonstrations of the need for future intervention."[75] Martin likens the shift toward a counterterrorism mode of US warfare—dispersed warfare fought by small groups of soldiers—to the logic of the financial arbitrageur who leverages volatility in risk markets for profit. Both demonstrate the temporality of the derivative: a present ruled by the promise of future instability.

For the US security state, climate change and terrorism both raise the threat level for unpredictable "shocks." Thus both have been the basis for a particular approach to risk management akin to what Walker and Cooper call a "culture of resilience"—an acceptance of perpetual flux in environmental and social "systems" and an adoption of security practices that foster "permanent adaptability in and through crisis."[76] We can see here a parallel between the US pivot to Africa and the ramping up of financialized agricultural development projects like ACRE and Pula. The logic undergirding both projects constructs Africa as a space of unending crisis that demands securitization. Resilience becomes the answer for individuals and governments seeking greater security. In the process, geographical peripheries of global agri-fintech markets and US warfare become productive frontiers for agribusiness, financial capital, and the expanding security state.

CONCLUSION: THINKING BEYOND AGRI-FINTECH

Promising a "lifeline" for smallholder farmers facing the devastation of the climate crisis, agri-fintech projects like ACRE and Pula continue to bring more farmers into financial and agricultural seed and input markets. As I have argued in this chapter, these efforts reproduce long-standing asymmetries of power that position particular people and places as permanently "at risk" while, at the same time, extracting wealth (here, in the form of pooled, financialized risk) primarily for the benefit of corporations in the Global North. Still, the agri-fintech promise remains a powerful discourse across industry and development sectors. Yet the dominant framing of the need to secure hundreds of millions of people on the front lines of climate crisis keeps us from asking a different set of questions about a more climate-just future. These questions would consider, for example, how alternative insurance mechanisms could be built on the premise that countries in the Global North owe countries in the Global South an "ecological debt." They would also ask how climate adaptation finance mechanisms might prioritize societal wellbeing over the profits of global financial institutions.

The agri-fintech securitizing project also aligns with the US security state's adoption of resilience thinking to anticipate and act upon a climate-changed future. Through this logic, US empire pivots to Africa as a geography of perennial crisis. More broadly, we can consider how the US empire extends to secure new frontiers produced through the climate crisis. But just like agri-fintech techno-optimism, a more resilient empire always expands through exclusion. Indeed, the work of empire is fundamentally about dividing the world into zones of security and zones waiting to be secured, the homeland and the frontier, green zones and red zones. As climate change drives social and political instabilities, the question of empire remains imperative for engaging with questions of justice. In this book's conclusion, I discuss the need to think beyond the security/securitizing paradigm of climate adaptation to imagine and enact a different future.

Conclusion

WHAT CAN THE GREEN REVOLUTION
TEACH US ABOUT CLIMATE CHANGE?

In a revealing moment at a press conference during the United Nations Convention on Climate Change International Meeting in 2021 (COP 26), US Speaker of the House, Nancy Pelosi, was asked about the relationship between American defense spending and the country's efforts to cut carbon emissions.[1] Reporter Abby Martin reminded Pelosi that she had just presided over yet another huge increase in the already-massive US military budget (by some estimates, now larger than the next nine countries combined). Martin also reminded Pelosi that the American military has a massive carbon footprint. "How," Martin asked, "can we seriously talk about net zero [carbon emissions] if there is this bipartisan consensus to constantly expand this large contributor to climate change?" Pelosi's answer was telling: "National security advisers all tell us that the climate crisis is a national security matter." In her hasty reply, Pelosi went on to repeat the word *security* three more times. She did not answer Martin's question. Of course, Pelosi was right, in the sense that the US government views climate change, first and foremost, through the lens of security. This point has been confirmed more recently through an interagency government report, in which different agencies associated with US National Defense reported on how the climate crisis might drastically reshape their everyday operations.² "Climate change is a national security matter" has become a standard line in American halls of power from the Pentagon to the White House.

I open this conclusion with Pelosi's comment because it suggests the broader political context in which the effects of climate change, especially but not only on the African continent, are perceived as a security issue. This book has shown how Green Revolution leaders from Norman Borlaug to Bill Gates cast the potential threat of hunger and poverty in the Global South as an "emergency" that threatens those who are seemingly sheltered from the worst of the storm in the Global North. From the imperialistic framings of the Green Revolution in Borlaug's era to warnings about food insecurity and climate change leading to "destabilized" regions today, security logics have advanced the "long" Green Revolution.³ Yet how might vantages other than security change how we conceptualize the politics of agricultural development? Though Gates and the United States continue to promote a view of global hunger that emphasizes security first, where might we turn for alternative framings? In what follows, I consider some general concluding points from the preceding chapters that might generate different ways for thinking about—and acting upon—global material inequalities that fuel disproportionate suffering and vulnerability.

1: THE DREAM OF DECONTEXTUALIZED SEED VS. AGROECOLOGY

One of the Monsanto officials I interviewed gave me an impromptu walking tour of the company's sprawling research facilities in the St. Louis suburb of Chesterfield. As we walked around the campus, some of which was under construction as part of a $400 million expansion project, I got a chance to see some of Monsanto's cutting-edge agricultural biotechnologies up close. My host took me down long halls lined with heavy steel doors that enclosed vault-like "growth chambers," in which Monsanto scientists could control every aspect of the growth environment. The cold silver doors were emblazoned with bright orange warning stickers announcing that the experiments contained genetically engineered plants. According to my informant, every aspect of the growing environment, from humidity to light to temperature to air pressure could be manipulated: "They can program it so it's exactly like the middle of Kenya."[4] I also got a chance to check out some of the company's in-house gene-sequencing machines, including a "seed chipper" that was described as the envy of their rivals because of how it allowed Monsanto scientists to "extract" plant DNA with the upmost precision. (The chipper, I was told, was the subject of "big litigation" around the question of whether Syngenta and DuPont's similar machines had infringed on Monsanto's patents.) With a clear dose of company pride, my host divulged that it was Monsanto's ability to extract, visualize, and then manipulate plant DNA that set the company apart.

My informant continued to elaborate on these newer gene-sequencing machines, some of which had been designed and built right there in Chesterfield. With them, the company was moving further away from being a traditional "seed" company and had begun to see itself as a "big data" company, finding it more efficient to "move fieldwork out of the fields and into the lab." Doing this work meant that the context from which the seed first grew mattered less and less. "It doesn't matter where the seed is located, because it just becomes information," my informant remarked. This encompasses Monsanto's shift from a seed company to a "data" company, a talking point that company officials were just beginning to use but that took on more meaning after Bayer acquired the company. Monsanto's "data" assets were a crucial part of the acquisition, but the fact that Bayer now controls large pools of data collected from Monsanto digital farming technologies has raised concerns among scholars and activists for the lack of "democratic institutional oversight" of the company's big data practices.[5] The local ecological and cultural context matters much less in this gene-as-data-as-property perspective. Along these lines, WEMA seeds moved along global circuits ("It's a global program," my informant reminded me): from trait "discovery" to integration to breeding, the project's seeds traveled across the company's global network, moving from Mexico to Hawai'i to St. Louis to South Africa to Kenya.

These two points that my Monsanto informant articulated—that genomic technology and big data practices had made the locality matter less and that, at the same time, the company utilized its own global networks as it developed its patented seeds—demonstrate how projects like Water Efficient Maize remove seeds from their local contexts. Indeed, WEMA officials often described the project's biotechnology as "scale neutral." The idea was that the GM seeds worked across scales. Bayer aims to expand on Monsanto's efforts to develop "scale-neutral" technologies to reach millions of smallholders. Speaking at a panel presentation at the 2023 World Economic Forum, Bayer's CEO Werner Baumann spoke about how the company has ambitious aims to tap the "huge potential" of the world's smallholder farmers. Baumann claimed they had a company goal of reaching one hundred million smallholders by 2030—and that they had already reached fifty million through "a lot of partnerships and, of course, technology."[6]

But this dream of decontextualized seeds renders local knowledge and local ecologies irrelevant. In this way, we might contrast the seeds of Big Biotech with a seed in context. This is precisely the approach that farmers and activists working under the banner of *agroecology* propose. The concept encapsulates a science and practice of agriculture that considers not only the unique environmental conditions of a given location, but also the cultural and knowledge systems that are part of that local environment. Maywa Montenegro de Wit summarizes scholars' and activists' approaches to agroecology, noting the expansive breadth of political and ecological approaches coalescing in and through what she calls the "science, practice, and movement."[7] As Montenegro de Wit explains, agroecology has been understood as "an emancipatory movement to increase farmers' power and control over their own production, as a pathway to revive Indigenous and traditional knowledge systems, and as a scientifically robust means of enhancing access to food grown in healthy, environmentally sound ways."[8] From these perspectives, agroecology emphasizes local knowledge and practices—and insists, too, on locating seeds and plants within local ecologies. In these ways, it is the opposite approach from the seed-as-data perspective my Monsanto informant detailed. Indeed, farmer and activist groups have used the concept in their efforts to resist the Green Revolution for Africa's agenda. The largest such group, the Alliance for Food Sovereignty in Africa, explicitly argues for agroecology as a foundation from which to grow more democratic and productive agricultural systems on the continent. As the organization's leaders wrote in an editorial denouncing the Gates Foundation's approach (memorably titled "Bill Gates Should Stop Telling Africans What Kind of Agriculture

CONCLUSION 131

Africans Need"), the organization views "agroecology as liberating—based on farmers' rights to choose seeds and methods of cultivation, and free of corporate interference and control."⁹ A growing number of scholars, activists, farmers, and policy advocates are mobilizing around agroecology as a political project that foregrounds questions of knowledge and power in discussions about the future of agriculture. *Seeding Empire* has excavated the driving logics behind the capitalist development paradigm agroecologists oppose. It has also offered historical context through which to better understand current debates over what food systems can and should be, especially in Africa.

2: "AFRICAN-LED" VERSUS "SOMEONE ELSE'S EVENT"

One of the central talking points in both AGRA and the WEMA/TELA projects is the idea that these Western-funded initiatives are "African-led." In my interviews with Monsanto and Gates Foundation officials, especially, this trope was often repeated. Two Monsanto officials that had worked on the WEMA project early on narrated the project in terms of its "origin" on the continent. "Remember," one scientist told me, "originally they [the AATF] were the ones who introduced the project and realized that 'we need to partner with Monsanto.'" Emphasizing this point, this same scientist later stated bluntly: "The origination of the project came from the African organizations."[10] When I asked about what role the Rockefeller Foundation, an American institution, had played, both scientists were unaware that the foundation helped establish these African organizations in the first place. Other Monsanto officials made it clear that the narrative of "African led," was part of how the company liked to depict the project: "We want this to be seen as an African-led project"—later in the conversation adding that the project was *"truly,* African led."[11] As I explained in chapter 4, Monsanto and other biotech companies had long been interested in using their proprietary biotech traits for projects aimed at reaching smallholder farmers in Africa. While it would be inaccurate to discount the fact that African-based scientists and politicians had also wanted to bring GM crops to their respective countries prior to WEMA, calling the project "African led" overlooks important details about how it emerged. At the same time, we should consider the narrative power embedded in the framing of transnational projects like WEMA as "African led."

This is not meant to suggest that there are not ways in which projects like WEMA/TELA are, indeed, driven from countries in Africa. At the same time, the global divisions in terms of who controls the property at the heart of biotechnology projects like WEMA/TELA is indisputable. As I showed in chapter 4, when genes are private property—when they are "someone else's event" in terms of intellectual property—they will always be under the control of their owners (in this case, Bayer). In most of the development projects around agricultural biotechnology in Africa, especially those facilitated through the AATF, the "someone" who owns that gene is one of the Big Three agricultural biotechnology companies: Bayer, Corteva, or Syngenta.

To what extent should projects based around a property regime and profit motive built around extracting wealth from the bottom of the value chain and moving it up for the benefit of Western shareholders be thought of as African-led? As I concluded in chapter 4, exactly which way the biotech revolution unfolds across sub-Saharan Africa is, in many ways, yet to be determined. And yet the commitments of the major actors I traced throughout the book, including the Gates Foundation, the US government, and the ag-biotechnology companies continue to promote the perspective that biotechnology projects will be essential for taking on food insecurity and poverty. American business lobby groups and government agencies like USAID and USDA continue to push African governments to deregulate the technologies. As an example, in 2022 AGRA partnered with the US Department of Agriculture to begin several projects aimed at bringing smallholder farmers into commercial seed markets. A key part of the agreement was, in the words of one industry group report, "to promote the adoption, application, and uptake of science and technologies."[12]

We can see further symbiosis between American development and agribusiness in a recent preferential trade agreement signed by the United States and Kenya. In the lead-up to the agreement, the US Chamber of Commerce (the largest business lobbying group in the United States) facilitated a meeting with the biotech multinationals to discuss how the potential trade agreement might catalyze more opportunities to deregulate biotechnologies.[13] Questions about how to develop Kenya's agriculture toward producing more imports for US markets were central to the meeting. ("How do we prepare the Agriculture sector for investment, knowledge, and technology transfer?") These are, of course, regulatory questions concerning how to move the government to deregulate technologies like biotech. When the trade agreement went into place in the summer of 2022, the US Office of Trade Representative press release identified the agricultural sector as a potential source of increased trade: "The two sides [Kenya and the United States] share an interest in . . . creating an enabling environment for innovative agricultural technologies that would help achieve food security goals." This surely includes agricultural biotechnologies. And the language of "good regulatory practices" indicates that making the regulatory environment more conducive to GM crops and, therefore, international trade will be a priority within the new trade agreement.[14] As I have pointed out throughout this book, it is important to situate the symbiotic relationship between US corporate interests (and, indeed, multinational corporations' interests, as well) and US "National Security" interests. Asking how US policies articulate with other developmental logics can yield crucial insights about the ongoing operation of the contemporary Green Revolution. This example of the US trade/national security nexus promoting the commercial interests of global multinationals leads me to my next concluding point, which asks us to question whether the profit motive is, after all, the best way to approach the problems of hunger and poverty.

3: IS THE PROFIT MOTIVE THE BEST WAY
TO GENERATE TECHNOLOGICAL INNOVATIONS?

For most of the people I interviewed for this book, the questions of who profits from agricultural development projects was less important than that of who benefits. The profit motive was widely viewed as a means toward the end of "helping smallholder farmers." Indeed, the language of "smart" business had been taken up across the networks of Green Revolution development institutions. As an example, one informant that had worked at the AATF and on USAID-funded biotech communications efforts talked about how they used to listen to people that spoke about the evils of companies like Monsanto. Now, they just saw them as practicing "smart business." As I wrote about in chapters 2 and 4, I would often pose questions to my informants along the lines of "But wait, isn't that going to line the pockets of the shareholders of these greedy agribusinesses?" Despite my (perhaps too) frequent pressing of this issue, most of my informants insisted that the interests of the farmers came first and the question of who profits was, if mentioned at all, mostly an afterthought. This not only demonstrates how the logic of philanthrocapital suffused through different levels of these projects, but also shows how corporate interests were often conflated with everyone's interests. Though representatives from the multinationals were candid that they could only do projects like WEMA because the Gates Foundation had removed their requirement to secure return-on-investment, officials from the Gates Foundation, USAID, and some of the public sector scientists seemed eager to provide the kind of initial investments that could lay the groundwork for future profit making. Again, my point is not to catch any of these noncorporate officials in a kind of "gotcha" moment—"see, you really *are* working for Monsanto!"—but to demonstrate the power of the philanthrocapitalist logic that the profit motive is the most efficient way to generate a socially beneficial "impact" that might benefit poor farmers.

But what if the profit motive really is not the best way to improve society? Jesse Goldstein's *Planetary Improvement* examines the limitations of the profit motive as a catalyst for innovation.[15] Through interviews with a range of workers working in the "green finance" sector, Goldstein shows that the need to generate return-on-investment led to a lack of innovation in "green tech." His conclusion offers a useful parallel for the world of agricultural development: contrary to popular mythology, capital can work to stifle innovation. Because it needs to generate money, first and foremost, capital is not always the best producer of social impact. Goldstein offers a pointed critique of the notion of "impact"—a concept also ubiquitous in the development and philanthrocapitalist world I researched—by showing how "impact" often translated to "impact as capital." In Green tech projects, making an impact often equated to, first and foremost, generating financial returns. What would happen, Goldstein asks, if we decoupled the pursuit of innovation and the profit motive? Goldstein's work makes a clear distinction that the issue at hand is not the technologies in the green tech sector, but the ways they are tethered tightly to the profit motive at every step. Extended to the world of ag-biotech, this shows us that the key point is not to dismiss the technology of biotech seeds (or, as is becoming increasingly common, the kinds of gene-edited seeds I discussed in chapter 3). Rather, we need to envision alternative socio-technical systems in which technological innovations including but not limited to biotech seeds might benefit more people.

This issue has only become more pressing with the consolidation of the ag-biotech industry and the related competition surrounding the patents involved in gene-editing techniques like CRISPR.[16] If it is any indication of the future political economic landscape, Corteva Agriscience is the largest holder of CRISPR patents. Millions have already been spent on legal fees and lawsuits. And billions will be reaped in licensing fees. Yet it is imperative to ask questions about how the private property regime curtails innovations or makes it more difficult to get potentially socially beneficial technologies into the hands of people who might benefit from them. Technologies like CRISPR open a range of possibilities—and are, perhaps unsurprisingly, highly touted by many of the same biotechnology proponents I discussed in this book. However, we must ask how these new technologies will extend the kind of exclusionary (and, arguably, monopolistic) situations that have defined the history of agricultural biotechnology.

4: A MATTER OF EMPIRE

"If Americans want to care about Africa, maybe they should consider evaluating American foreign policy." These words are drawn from a frequently cited online essay, in which the writer Teju Cole excoriated what he called the white-savior industrial complex. Cole wrote that humanitarian projects reproduce tired tropes about suffering Africans needing to be rescued. Cole and others' critiques of the "white savior" has brought the term into mainstream conversations. But one part of Cole's critique has perhaps received less attention: his call to pay attention to the impact of US empire on the continent's politics. For Americans, in particular, this entails examining the roots of American exceptionalism and reconsidering the historical and geographical connections between "the West and the Rest."

One of this book's central lessons is, to paraphrase Cole: if you care about people suffering around the world due to human-induced climate change, ask first about the workings of your own government.

In the debates over whether African countries should approve GM crops, Western journalists and biotech proponents rehearse the argument that the debate over what technologies should be used to "help" farmers in Africa is a discussion that only privileged Northerners can afford to have. In the words of a *Washington Post* journalist: "The last thing Africa needs to be debating is GMOs."[17] How dare liberal Westerners argue over the merits of a technology that could help impoverished Africans! This was the same line of argument that Bill Gates picked up on in his World Food Prize speech that I quoted at the beginning of the book. Gates drew raucous applause from the Des Moines crowd for chastising biotech critics that failed to appreciate the "emergency" facing farmers in Africa. Yet, as Cole reminds us, the will to save also functions as a kind of "release valve" that lets out some of the pressure built up through an altogether exploitive and violent system. In sharp words, Cole reminds those in the United States who feel sorry for people facing violence in Africa to think more critically about the history of US foreign policy and its ties to a rapacious racial capitalism. "The White Savior Industrial Complex," Cole reminds us, "is a valve for releasing the unbearable pressures that build in a system built on pillage. We can participate in the economic destruction of Haiti over long years, but when the earthquake strikes it feels good to send $10 each to the rescue fund."

Cole's admonishing of liberal do-gooders that want to "uplift" Africans without confronting the history of their own government's role in destabilizing conditions across the continent and elsewhere is suggestive of another of *Seeding Empire*'s main points: though conversations about the future of food on the African continent might lead to strong debates about what the "right thing to do" is (as my Monsanto informants used to describe WEMA), before we in the United States project any sort of liberal fantasy onto the continent or its people, we need to take a much more careful look into the mirror. Journalism and scholarship on the US military's growing, though mostly unacknowledged, footprint across the continent is an apt place to start. *Seeding Empire* adds to this conversation through tracing an emerging financialized agricultural development/national security nexus.

The changing landscape around financial development and climate adaptation projects and US foreign policy that I analyzed in chapter 5 offers us ways to further complicate narratives of US-Africa "partnerships." Understanding the connections between trade and security is also an important line of inquiry for considering the changing priorities of US geopolitics in connection to Africa. At the time of this writing, the Biden administration has recently released its official "U.S. Strategy toward Sub-Saharan Africa."[18] Though the document follows much of the arguments of the Obama administration's approach to US policy across the continent, it is notable that it argues for increased security and trade efforts. As *Foreign Policy* reported, "The United States will now seek to double down on its enormous soft-power influence across the continent by leveraging U.S. private sector trade and investment, which it has previously underutilized."[19] As recent signs point to the United States continuing to ramp up its military investments on the continent, the strategy is framed in terms of a new developing Cold War–type great power struggle. (A framing that took on even greater importance after Russia's 2022 invasion of Ukraine, after which a substantial number of African nations refused to side with the United States and condemn Russia through United Nations sanctions.) The strategy frames both Russia and China in terms of direct threats to US interests on the continent. In several ways, it extends the Obama administration's emphasis on security framing in terms of the global war on terror.[20] To grasp the implications of the trade/security nexus, we must take a longer-term view and consider how the projection of "soft," developmental power is integral to an always-in-process American empire. More recently, this power includes US trade interests partnering with security regimes in African countries. As I have shown throughout this book, understanding how US empire articulates with other interests and logics—be it philanthrocapital or development—can clarify the contemporary contours of US power.

5: "SO WHAT DOES NORMAN BORLAUG HAVE TO DO WITH CLIMATE CHANGE?"

For my fifth, and final, closing point, I want to return to the two figures with whom I began this book: Borlaug and Gates—the "father of the Green

Revolution" and his wealthiest follower. As I discussed in the book's first two chapters, it was Borlaug's story that first inspired Gates to take up the cause of feeding impoverished farmers with Western technologies like biotech crops. The billionaire philanthropist has become fond of telling the Borlaug story when he explains his foundation's approach to agriculture (which, importantly, informs how it spends billions of dollars—over six billion (USD) between 2003 and 2020, according to one study.)[21] In his 2021 book, *How to Avoid a Climate Disaster: The Solutions We Have and the Breakthroughs We Need,* for example, Gates begins a chapter on agriculture by recounting the Borlaug narrative. Like Borlaug, Gates begins with the specter of population growth. He cites Paul Ehrlich's warnings from *The Population Bomb* (1968) of impending famines. Yet Gates argues that Ehrlich and the other "doomsayers" got one thing wrong. They failed to consider "the power of innovation." "They didn't account for people like Norman Borlaug, the brilliant plant scientist who sparked a revolution in agriculture," Gates asserts. Detailing the Borlaug hero narrative, Gates summarizes: "Starvation plummeted, and today Borlaug is widely credited with saving a billion lives."[22] Gates goes on to suggest that Borlaug offers a clear example of the kind of scientific mind we need today to take on the twin challenges of climate change and population growth.

Tellingly, Gates portrays Borlaug as an "innovator," rather than a unique kind of scientist-statesman who pushed governments to adopt US and agribusiness-friendly policies. Gates removes Borlaug from the geopolitics of his day and simply has him serve as a kind of "groundbreaking," singularly "brilliant" person that launched the Green Revolution. Perhaps unsurprisingly, Gates leaves aside the massive level of state support that buttressed the Green Revolution, on the part of both the United States and the countries that Borlaug pushed into adopting his new varieties.[23] Yet the memory of Borlaug as a kind of technocratic "wizard" has prevailed.[24] As I argued in chapter 1, this has much to do with the power of the Borlaug myth to both tap into dominant cultural memories and, in turn, to shape those memories. Both Gates and Borlaug provide useful examples of the persistent power of White ignorance. Their technological prophecies teach lessons about global poverty that cultivate a worldview that fails to recognize the historical roots of contemporary inequalities. More to the point: they instruct audiences to read the world in a way that cannot see how "rich world" wealth feeds upon and creates "poor world" immiseration.

Like Gates, I also consider it crucial that we ask what Norman Borlaug can teach us about climate change. However, as I have shown throughout this book, the lessons we should draw from the story of Borlaug and the Green Revolution are quite different from those Gates offers. Both Borlaug and Gates reproduce a form of relating to the past that disavows histories of racial capitalism's uneven development and the violence of slavery, settler colonialism, and overseas empire. Like Borlaug, Gates teaches audiences to think about global poverty through a view that separates those who "suffer most" from what Gates dubs the "rich world."[25] This way of understanding global geography actively refuses a knowledge of the ways in which the plight of the global rich and poor have been mutually formed. As *Seeding Empire* has shown, the lesson we should learn from the ongoing Green Revolution is that histories of dispossession and uneven development continue to shape the lives of farmers around the world. As I have suggested, the Borlaug story might spur us to raise a different set of questions about the relationship between what Borlaug called the "privileged" and "forgotten" worlds. How would international debates around climate policy shift if countries in the Global North took seriously the climate debt they owed to the people of the Global South? How might we extend the climate conversation beyond questions of technological innovation and move toward the difficult, yet necessary questions about responsibility and repair? How might we develop approaches to climate adaptation that privilege societal progress rather than the profits of global agribusiness and financial corporations? And how might we insist upon a different way of remembering the Green Revolution's past—and, in turn, charting our collective future?

Notes

INTRODUCTION

1. Bill Gates, "Support for the World's Poorest Farmers" (World Food Prize, October 15, 2009), https://www.worldfoodprize.org/documents/filelibrary/images/borlaug_dialogue/2009_speakers/transcripts/2009BorlaugDialogueGatesbrief_65B2AF6BB5B25.pdf. During the 2017 World Food Prize, the event's MC, former US ambassador and current director of the World Food Prize Foundation, Kenneth Quinn told the conference attendees that before Gates's 2009 appearance in Des Moines, then director of the Gates Foundation's Agricultural Development program, Raj Shah, had called Quinn on the phone and insisted that Gates wanted to announce his "big initiative for Africa" at the annual conference. Author field notes.

2. Borlaug founded the World Food Prize in 1986, with the stated intention of developing a kind of Nobel Prize for food and agriculture. The $250,000 annual award is given each year during a ceremony at the Iowa State Capitol. Leon F. Hesser, *The Man Who Fed the World: Nobel Peace Prize Laureate Norman Borlaug and His Battle to End World Hunger: An Authorized Biography* (Dallas, TX: Durban House, 2006), 137.

3. Tom Philpott, "Bill Gates Reveals Support for GMO Ag," *Grist*, October 22, 2009, https://grist.org/article/2009-10-21-bill-gates-reveals-support-for-gmo-ag/.

4. Robert L. Paarlberg, *Starved for Science: How Biotechnology Is Being Kept Out of Africa* (Cambridge, MA: Harvard University Press, 2009). For an overview of Monsanto's long-standing, but decidedly less-than-successful efforts to brand its GM crops "pro-poor," see Dominic Glover, "The Corporate Shaping of GM Crops as a Technology for the Poor," *Journal of Peasant Studies* 37, no. 1 (January 2010): 67–90, https://doi.org/10.1080/03066150903498754.

5. Ambassador Quinn recalled Gates's speech prompting an unprecedented ovation during the 2016 World Food Prize. In 2017, Quinn noted that Gates's address had been *the* moment of his nearly twenty years hosting the conference. Author field notes.

6. My use of *genealogy* follows Michel Foucault, who memorably called for studying the history of ideas without assumptions about progress or inevitability. See: Michel Foucault, "Nietzsche, Genealogy, History." In *The Foucault Reader*, edited by Paul Rabinow, 76–100. New York: Pantheon, 1984. Geographer Vinay Gidwani defines this approach as "a geography and a history . . . where analysis proceeds not from the certitude of given categories but instead takes as its philosophical task to ask how those categories acquired their givenness and with what consequences." Vinay K. Gidwani, *Capital, Interrupted: Agrarian Development and the Politics of Work in India* (Minneapolis: University of Minnesota Press, 2008), xvii.

7. Raj Patel, "The Long Green Revolution," *Journal of Peasant Studies* 40, no. 1 (January 2013): 1–63, https://doi.org/10.1080/03066150.2012.719224.

8. Nick Cullather, *The Hungry World: America's Cold War Battle against Poverty in Asia* (Cambridge, MA: Harvard University Press, 2013), chap. 10.

9. Jamey Essex, *Development, Security, and Aid: Geopolitics and Geoeconomics at the U.S. Agency for International Development* (Athens: University of Georgia Press, 2013).

10. Garrett Graddy-Lovelace, "The Coloniality of US Agricultural Policy: Articulating Agrarian (in)Justice," *Journal of Peasant Studies* 44, no. 1 (January 2, 2017): 78–99, https://doi.org/10.1080/03066150.2016.1192133. Other scholarship that has influenced my thinking about agriculture and empire include: David A. Chang, *The Color of the Land: Race, Nation, and the Politics of Landownership in Oklahoma, 1832–1929*, (Chapel Hill: University of North Carolina Press, 2010); Frieda Knobloch, *The Culture of Wilderness* (Chapel Hill: University of North Carolina Press, 1996); Hannah Holleman, *Dust Bowls of Empire: Imperialism, Environmental Politics, and the Injustice of "Green" Capitalism* (New Haven, CT: Yale University Press, 2018). The description of empire as "always in process" comes from the introduction to a recent special issue on empire in *American Quarterly*, Christopher Lee and Melani McAlister, "Introduction: Generations of Empire in American Studies," *American Quarterly* 74, no. 3 (2022): 477–97, https://doi.org/10.1353/aq.2022.0031.

11. Books that have been particularly useful in helping me frame questions around empire and history include: Jodi A. Byrd, *The Transit of Empire:*

Indigenous Critiques of Colonialism (Minneapolis: University of Minnesota Press, 2011); Lisa Lowe, *The Intimacies of Four Continents* (Durham, NC: Duke University Press, 2015); Moon-Ho Jung, *Menace to Empire: Anticolonial Solidarities and the Transpacific Origins of the US Security State* (Oakland: University of California Press, 2023); Megan Black, *The Global Interior: Mineral Frontiers and American Power* (Cambridge, MA: Harvard University Press, 2018); Daniel Immerwahr, *How to Hide an Empire: A History of the Greater United States*, reprint ed. (New York: Picador, 2020); Amy Kaplan and Donald E. Pease, *Cultures of United States Imperialism* (Durham, NC: Duke University Press, 1993).

12. "Remarks by the President at Symposium on Global Agriculture and Food Security," White House, May 18, 2012, https://obamawhitehouse.archives.gov/the-press-office/2012/05/18/remarks-president-symposium-global-agriculture-and-food-security.

13. "Project archive" is a term from Philip Joseph Deloria and Alexander I. Olson, *American Studies: A User's Guide* (Oakland: University of California Press, 2017).

14. One of the archivists I worked with told me about the Rockefeller Foundation's internal culture of archiving everything.

15. James Sumberg, Dennis Keeney, and Benedict Dempsey, "Public Agronomy: Norman Borlaug as 'Brand Hero' for the Green Revolution," *Journal of Development Studies* 48, no. 11 (November 2012): 1587–600, https://doi.org/10.1080/00220388.2012.713470.

16. I also utilized the records of E. C. Stakman, Borlaug's PhD advisor and a senior advisor to the Rockefeller Foundation during the formative years of their early Green Revolution projects. As a member of the foundation's "Board of Agricultural Consultants," which charted the overall direction and recommended funding priorities for its international agricultural program, Stakman was a key figure in shaping Rockefeller Foundation programs.

17. A social scientist I met was refused interviews after their work was deemed too politically charged.

18. Bartow J. Elmore, *Seed Money: Monsanto's Past and Our Food Future* (New York: W.W. Norton & Company, 2021).

19. I attended the annual conference each year between 2014 and 2017.

20. Laura Nader, "Up the Anthropologist: Perspectives Gained from Studying Up," in *Reinventing Anthropology*, ed. Dell Hymes (New York: Random House, 1969).

21. William G. Moseley and Melanie Ouedraogo, "When Agronomy Flirts with Markets, Gender, and Nutrition: A Political Ecology of the New Green Revolution for Africa and Women's Food Security in Burkina Faso," *African Studies Review* 65, no. 1 (March 2022): 41–65, https://doi.org/10.1017/asr.2021.74; Jessie K. Luna, "The Chain of Exploitation: Intersectional Inequalities, Capital Accumulation, and Resistance in Burkina Faso's Cotton Sector," *Journal of Peasant Studies* 46, no. 7 (November 10, 2019): 1413–34, https://doi.org/10.1080/03066150

.2018.1499623; Joeva Sean Rock, *We Are Not Starving: The Struggle for Food Sovereignty in Ghana* (Lansing: Michigan State University Press, 2022); Matthew A. Schnurr, "Biotechnology and Bio-Hegemony in Uganda: Unraveling the Social Relations Underpinning the Promotion of Genetically Modified Crops into New African Markets," *Journal of Peasant Studies* 40, no. 4 (July 2013): 639–58, https://doi.org/10.1080/03066150.2013.814106.

CHAPTER 1. HOW WE REMEMBER THE GREEN REVOLUTION

1. James Sumberg, Dennis Keeney, and Benedict Dempsey, "Public Agronomy: Norman Borlaug as 'Brand Hero' for the Green Revolution," *Journal of Development Studies* 48, no. 11 (November 2012): 1587–600, https://doi.org/10.1080/00220388.2012.713470.

2. Marita Sturken, *Tangled Memories: The Vietnam War, the Aids Epidemic, and the Politics of Remembering* (Berkeley: University of California Press, 1997).

3. When the long-running PBS documentary program, *American Experience* tried to balance some of its celebration of Borlaug's legacy with a discussion of the Green Revolution's "unintended consequences," it sparked an angry response in a *Wall Street Journal* editorial that admonished the filmmakers for failing to state unequivocally that Borlaug was a hero for all. (Though, as I discuss later in the chapter, the *American Experience* episode mostly celebrated Borlaug.) "Battering Norman Borlaug," *Wall Street Journal* (Eastern Edition), April 25, 2020.

4. Useful assessments of Borlaug's work include: Marci Baranski, *The Globalization of Wheat: A Critical History of the Green Revolution* (Pittsburgh, PA: University of Pittsburgh Press, 2022); Glenn Davis Stone, *The Agricultural Dilemma: How Not to Feed the World* (New York: Routledge, 2022); Tore C. Olsson, *Agrarian Crossings: Reformers and the Remaking of the US and Mexican Countryside* (Princeton, NJ: Princeton University Press, 2017); Nick Cullather, *The Hungry World: America's Cold War Battle against Poverty in Asia* (Cambridge, MA: Harvard University Press, 2013). Gabriela Soto Laveaga, "Beyond Borlaug's Shadow: Octavio Paz, Indian Farmers, and the Challenge of Narrating the Green Revolution," *Agricultural History* 95, no. 4 (October 1, 2021): 576–608, https://doi.org/10.3098/ah.2021.095.4.576.

5. Kenneth Quinn, "Extended Biography: Norman E. Borlaug," 2009, https://www.worldfoodprize.org/en/dr_norman_e_borlaug/extended_biography/.

6. Quinn.

7. Roxanne Dunbar-Ortiz, *An Indigenous Peoples' History of the United States* (Boston: Beacon Press, 2014), xxi.

8. On settler colonialism as a process, see: Rita Dhamoon, "A Feminist Approach to Decolonizing Anti-Racism: Rethinking Transnationalism, Intersec-

tionality, and Settler Colonialism," *Feral Feminisms* 4 (Summer 2015), https://feralfeminisms.com/rita-dhamoon/; Kevin Bruyneel, *Settler Memory : The Disavowal of Indigeneity and the Politics of Race in the United States* (Chapel Hill: University of North Carolina Press, 2021). The classic description of settler colonialism as a structure comes from Patrick Wolfe, "Settler Colonialism and the Elimination of the Native," *Journal of Genocide Research* 8, no. 4 (December 2006): 387–409, https://doi.org/10.1080/14623520601056240.

9. Dunbar-Ortiz, *An Indigenous Peoples' History*, 5.

10. Charles C. Mann, *The Wizard and the Prophet: Two Remarkable Scientists and Their Dueling Visions to Shape Tomorrow's World* (New York: Vintage, 2018).

11. David Vine, *The United States of War: A Global History of America's Endless Conflicts, from Columbus to the Islamic State* (Oakland: University of California Press, 2020), 94.

12. Mann, *The Wizard and the Prophet*, 97.

13. Bruyneel, *Settler Memory*.

14. "In a settler society," Bruyneel writes, "the work of collective memory serves to reaffirm the settler claim of belonging to, appropriation of, and authority over lands, on the one hand, and the disavowal of the genocide, dispossession, and alienation of Indigenous peoples, on the other hand." Bruyneel, 14.

15. Amy Kaplan, "'Left Alone with America': The Absence of Empire in the Study of American Culture," in *Cultures of United States Imperialism*, ed. Amy Kaplan and Donald E Pease (Durham, NC: Duke University Press, 1993).

16. Bruyneel, *Settler Memory*, 12.

17. Quoting Borlaug's Mexican Agriculture Program colleagues, biographer Leon Hesser describes Borlaug's "fanatical devotion to wheat." Hesser, *The Man Who Fed the World*.

18. CIMMYT, "Norman Borlaug Man of the Year 1969," YouTube, May 2011, https://www.youtube.com/watch?reload=9&v=r-8HWsfntEQ.

19. These tropes are ubiquitous in agribusiness and development circles, and extend to popular press coverage of Borlaug as well. The front-page *New York Times* obituary noted that Borlaug was "widely described as the father of the broad agricultural movement called the Green Revolution." The *Times* described Borlaug as "the plant scientist who did more than anyone else in the 20th century to teach the world to feed itself . . . whose work was credited with saving hundreds of millions of lives." See: Justin Gillis, "Norman Borlaug, Father of a Crop Revolution, Dies at 95," *New York Times*, September 13, 2009, sec. Energy & Environment, https://www.nytimes.com/2009/09/14/business/energy-environment/14borlaug.html.; At the 2014 unveiling of a Borlaug statue in the US Capitol, House Speaker John Boehner quipped that "it will be awfully nice to have a miracle worker around here." "Norman Borlaug Statue Unveiled at US Capitol," accessed May 15, 2019, https://www.mprnews.org/story/2014/03/25

/news/borlaug-statue; This "miracle worker" trope has even appeared in popular culture references to Borlaug. As Nick Cullather details, a 2000 episode of the popular primetime TV show, *The West Wing*, featured the US president (played by Martin Sheen) explaining at a press conference: "There are people who make miracles in the world. One of them lives right here in the U.S."—before recounting the popular narrative of Borlaug "saving India" from famine with his miracle wheat. See Nick Cullather, "Miracles of Modernization: The Green Revolution and the Apotheosis of Technology," *Diplomatic History* 28, no. 2 (2004): 227–54. For a general overview (and critique) of the Borlaug hagiography, see Sumberg et al., "Norman Borlaug as 'Brand Hero.'"

20. In *American Prophecy: Race and Redemption in American Political Culture*, George Shulman writes, "The office of prophecy is a public vocation mediating between a community and powerful realities it does not understand or control. In each regard, prophets make claims about the circumstances and difficulties—and fateful decisions—*of the whole*; indeed, in this way they reconstitute the very 'we' they seem to invoke as a given. In each regard, they seek to *redeem* the community they address and whose fate they commit to sharing. Prophecy is thus a performance to incite audiences to self-reflection and action. Not only a rhetorical act, prophecy is an embodied form of symbolic action." George M. Shulman, *American Prophecy: Race and Redemption in American Political Culture* (Minneapolis: University of Minnesota Press, 2008), 6.

21. As an example, the *New York Times* obituary connected Borlaug's remarkable skills as a wheat breeder to his "farm boy's instinctive feel for the plants and the soil from which they grew." Gillis, "Norman Borlaug."

22. Timothy A. Wise, *Eating Tomorrow: Agribusiness, Family Farmers, and the Battle for the Future of Food* (New York: The New Press, 2019), 111–13.

23. Stone, *The Agricultural Dilemma*, 46.

24. CIMMYT, "Norman Borlaug Man of the Year 1969."

25. Theodore W. Schultz, *Transforming Traditional Agriculture* (New Haven, CT: Yale University Press, 1964).

26. David Theo Goldberg, *Are We All Postracial Yet?* (Malden, MA: Polity Press, 2015).

27. Michael Adas, *Machines as the Measure of Men: Science, Technology, and Ideologies of Western Dominance* (Ithaca, NY: Cornell University Press, 2014).

28. Jean M. O'Brien, *Firsting and Lasting: Writing Indians out of Existence in New England* (Minneapolis: University of Minnesota Press, 2010); Bruyneel, *Settler Memory*.

29. The film cuts from Borlaug warning that humankind might soon go the way of the dinosaurs to stock photographs of famished children, giving British audiences a moral connection between hunger in the abstract and images of abject Black and brown bodies. See James Vernon for a discussion of the historical context of Western famine imagery and photography. James Vernon, *Hunger: A Modern History* (Cambridge, MA: Belknap Press, 2007).

30. Glenn Davis Stone walks through the empirical challenges to Malthus's argument in his recent book. Stone, *The Agricultural Dilemma*.

31. Stuart Hall, "The West and the Rest: Discourse and Power," in *Modernity: An Introduction to Modern Societies*, ed. Stuart Hall et al. (Malden, MA: Blackwell, 1996), 184–227.

32. Nicholas Hildyard, "'Scarcity' as Political Strategy: Reflections on Three Hanging Children," in *The Limits to Scarcity*, ed. Lyla Mehta (London: Earthscan, 2010), 161 n30.

33. Kalpana Wilson, *Race, Racism and Development: Interrogating History, Discourse and Practice* (London: Zed Books, 2012).

34. For more on the persistence of Malthusian thinking in Western development theory and policy, see: Eric B. Ross, *The Malthus Factor: Population, Poverty, and Politics in Capitalist Development* (London: Zed Books, 1998); and Lyla Mehta, ed., *The Limits to Scarcity: Contesting the Politics of Allocation* (London: Earthscan, 2010).

35. Stone, *The Agricultural Dilemma*.

36. This imagery built directly on the warnings of Neo-Malthusians such as Paul Ehrlich, whose bestselling 1968 book famously opened with a depiction of a "crowded slum area" in Delhi, where Ehrlich and his wife are "frightened" by numerous people all around them. "The streets seemed alive with people. People eating, people washing, people sleeping. People visiting, arguing, and screaming. People thrusting their hands through the taxi window, begging. People defecating and urinating. People clinging to buses. People herding animals. People, people, people, people." Ehrlich, *The Population Bomb*, 1.

37. Ross, *The Malthus Factor*.

38. Mytheli Sreenivas, *Reproductive Politics and the Making of Modern India* (Seattle: University of Washington Press, 2021).

39. Goldberg, *Are We All Postracial Yet?*

40. Wilson, *Race, Racism and Development*, 79.

41. Borlaug served as the director for two organizations advocating for population control. He served as the board of director for a third. See his CV: "Dr. Borlaug's CV," World Food Prize, accessed May 15, 2019, https://www.worldfoodprize.org/index.cfm?nodeID=87450&audienceID=1. For broader historical context, see: Matthew Connelly, *Fatal Misconception: The Struggle to Control World Population* (Cambridge, MA: Harvard University Press, 2010).

42. Musab Younis, "To Own Whiteness," *London Review of Books*, February 10, 2022, https://www.lrb.co.uk/the-paper/v44/n03/musab-younis/to-own-whiteness.

43. Richard Dyer, *White* (London: Routledge, 1997), 2.

44. Charles W. Mills, "White Ignorance," in *Race and Epistemologies of Ignorance*, ed. Shannon Sullivan and Nancy Tuana (Albany: State University of New York Press, 2007), 13–38.

45. Walter Rodney, *How Europe Underdeveloped Africa* (Washington, DC: Howard University Press, 1981).

46. Sumberg, et al., "Public Agronomy." For a broader discussion of DDT, see David Kinkela, *DDT and the American Century: Global Health, Environmental Politics, and the Pesticide That Changed the World* (Chapel Hill: University of North Carolina Press, 2013).

47. Norman Borlaug, "Mankind and Civilization at Another Crossroad" (FAO, Rome, Italy, November 8, 1971), https://www.fao.org/3/c3017e/c3017e.pdf.

48. Norman Borlaug, "Ending World Hunger: The Promise of Biotechnology and the Threat of Antiscience Zealotry," *Plant Physiology* 124, no. 2 (2000): 487–90.

49. Borlaug, 490.

50. Norman E. Borlaug, "Mobilizing Science and Technology for a Green Revolution in Achieving Greater Impact from Research Investments in Africa," ed. Steven Breth (Mexico City: Sasakawa Africa Association, 1996), 209–17.

51. Borlaug, 217.

52. Norman E. Borlaug, "A Green Revolution for Africa," *Wall Street Journal*, October 26, 2007, sec. Opinion, https://www.wsj.com/articles/SB119336762148772617; Norman Borlaug, "Feeding a Hungry World," *Science* 318, no. 5849 (October 19, 2007): 359, https://doi.org/10.1126/science.1151062.

CHAPTER 2. "A GREEN REVOLUTION, THIS TIME FOR AFRICA"

1. "WFP Founder Norman Borlaug Receives America's Highest Civilian Honor," World Food Prize, accessed May 14, 2019. https://www.worldfoodprize.org/index.cfm/87428/40024/wfp_founder_norman_borlaug_receives_americas_highest_civilian_honor. This chapter borrows its title from that of a 2014 *New York Times* opinion column. Part of their "solutions journalism" series, the article used the Borlaug hero story to call for a Green Revolution in Africa. See: Tina Rosenberg, "A Green Revolution, This Time for Africa," *New York Times*, Opinionater, April 9, 2014. https://archive.nytimes.com/opinionator.blogs.nytimes.com/2014/04/09/a-green-revolution-this-time-for-africa/.

2. The younger Borlaug describes assuring her dying grandfather that he had already done enough to catalyze a Green Revolution in Africa—through training and mentoring "thousands" of people that would continue his legacy. She assured him that his legacy would continue and that he had already "inspired" so many people "in Africa" to "take up the charge." See: Geoffrey Onditi, "Saturday Morning Interview (Julie Borlaug) KBC," YouTube, September 19, 2016, https://www.youtube.com/watch?v=VoFZWofCfEs.

3. Indeed, the final words and dying regret are often repeated together. Kenneth Quinn, the former US ambassador to Cambodia and president of the World Food Prize Foundation since 2000, tells a slightly different version of Borlaug's deathbed declarations than Julie Borlaug. In a 2013 tribute to Borlaug, Quinn writes: "Dr. Borlaug's last words were 'Take it to the farmer.' Just before that,

he said, 'I have a problem: Africa,' referring to his unfulfilled goal of bringing enhanced agricultural production to that continent." Just as in Julie Borlaug's version of the story, the combination of Borlaug's failure or "problem," with his commandment reinforces the position of Africa as that of being "not-yet" redeemed by the hero scientist. Kenneth Quinn, "Quinn: A Tribute to Norman Borlaug on the Fourth Anniversary of His Death," https://www.worldfoodprize.org/index.cfm/87428/40197/quinn_a_tribute_to_norman_borlaug_on_the_fourth_anniversary_of_his_death.

4. John Kerry repeated the narrative to African heads of state at a US-Africa summit in Washington. D.C., in 2014. See: John Kerry, "Remarks at a Working Session on Resilience and Food Security in a Changing Climate," US State Department, August 4, 2014, https://20092017.state.gov/secretary/remarks/2014/08/230219.htm. At the 2016 World Food Prize Conference (WFP), World Bank President Jim Yong Kim stressed that Borlaug had been unique in his commitment to applying the tools of science to pressing humanitarian concerns. At the 2017 WFP, African Development Bank president and World Food Prize laureate Akinwumi Ayodeji Adesina emphasized Borlaug's failed mission in Africa. Agribusiness CEOs from companies like Bayer and DowDuPont frequently invoke the Borlaug story—and Borlaug's commandment to "take it to the farmer"—in calls to expand seed and biotechnology markets across Africa. WFP citations are from author field notes and transcripts of conference program, available at https://www.worldfoodprize.org/index.cfm?nodeID=87431&audienceID=1.

5. Nick Cullather, *The Hungry World: America's Cold War Battle against Poverty in Asia* (Cambridge, MA: Harvard University Press, 2013).

6. "I have learned what's possible in agriculture from studying Borlaug and what's happened in the decades since his breakthroughs," Gates writes. Bill Gates, "Helping Poor Farmers Grow Their Crops," *GatesNotes* (blog), January 24, 2012, https://www.gatesnotes.com/The-Man-Who-Fed-the-World.

7. The Gateses have cultivated a branded identity around being "impatient optimists." They self-referentially use the phrase when describing their approach to philanthropy. The phrase is central to their foundation's brand as well, including its "impatient optimists" blog and Janet Echelman's aerial sculpture connecting the foundation's two headquarters buildings in Seattle, which is called "impatient optimist." Lisa Rogak's 2012 book on Bill Gates "in his own words" is likewise entitled *The Impatient Optimist*.

8. Ananya Roy, "Introduction: The Aporias of Poverty," In *Territories of Poverty*, ed. Ananya Roy and Emma Shaw Crane (Athens: University of Georgia Press, 2015), 1–36.

9. Karl Marx, *Capital: A Critique of Political Economy, Volume 1*, trans. Ben Fowkes, reprint ed. (London: Penguin Classics, 1992), chap. 10.

10. Author interview, Bill and Melinda Gates Foundation, Seattle, WA, July 25, 2023.

11. Rachel Schurman, "Micro(Soft) Managing a 'Green Revolution' for Africa: The New Donor Culture and International Agricultural Development," *World Development* 112 (December 2018): 180–92, https://doi.org/10.1016/j.worlddev.2018.08.003.

12. Linsey McGoey, *No Such Thing as a Free Gift: The Gates Foundation and the Price of Philanthropy* (London: Verso, 2016).

13. Hannah Appel, *The Licit Life of Capitalism: US Oil in Equatorial Guinea* (Durham NC: Duke University Press, 2019), 28.

14. The Rockefeller Foundation, "Africa's Turn: A New Green Revolution for the 21st Century," July 2006, https://assets.rockefellerfoundation.org/app/uploads/20060701123216/dc8aefda-bc49-4246-9e92-9026bc0eed04-africas_turn.pdf.

15. Matthew A. Schnurr, *Africa's Gene Revolution: Genetically Modified Crops and the Future of African Agriculture* (Montreal: McGill-Queen's University Press, 2019), 12.

16. Justin Gillis, "Can the Yield Gap Be Closed—Sustainably?," *New York Times* Green Blog, June 7, 2011, https://www.proquest.com/docview/2216870574/abstract/1882F1BC64ED4686PQ/1.

17. Bill Gates, *How to Avoid a Climate Disaster: The Solutions We Have and the Breakthroughs We Need* (New York: Knopf, 2021), 119.

18. Schnurr, *Africa's Gene Revolution*, 195.

19. "U.S. Seed Industry Has a Role to Play in Sub-Saharan Africa," June 21, 2011, http://www.amseed.org.

20. Jack Ralph Kloppenburg Jr., *First the Seed: The Political Ecology of Plant Biotechnology, 1492–2000.* 2nd ed. (Madison: University of Wisconsin Press, 2004).

21. Author interview, agricultural scientist, Des Moines, Iowa, October 14, 2015.

22. Calestous Juma, "How to Improve Africa's Seed Industry," World Economic Forum, September 11, 2015, https://www.weforum.org/agenda/2015/09/how-to-improve-africas-seed-industry/.

23. Alliance for a Green Revolution in Africa, *Seeding an African Green Revolution: The PASS Journey* (Nairobi, Kenya, 2017).

24. Alliance for a Green Revolution in Africa, 62.

25. Mark Kinver, "Lack of Seeds Hampers Africa's Ability to Boost Yields," BBC News, March 10, 2016, sec. Science & Environment, https://www.bbc.com/news/science-environment-35774445.

26. Author interview, Bill and Melinda Gates Foundation, Seattle, WA, July 22, 2015.

27. Alan Bjerga, "Sowing the Seeds of a Farm Boom in Africa," *Bloomberg*, March 31, 2016, https://www.bloomberg.com/news/articles/2016-03-31/sowing-the-seeds-of-a-farm-boom-in-africa.

28. Phil Howard's research and visual depictions of seed industry consolidation are indispensable. See the recent blog post, for example: Philip Howard, "Recent Changes in the Global Seed Industry and Digital Agriculture Industries,"

Philip H. Howard.Net (blog), January 4, 2023, https://philhoward.net/2023/01/04/seed-digital/.

29. Bob Koigi, "Corteva Agriscience Launches Operations in East Africa," Africa Business Communities, March 9, 2019, https://africabusinesscommunities.com/news/corteva-agriscience-launches-operations-in-east-africa/.

30. Phil Howard, "Global Seed Industry Changes since 2013," *Philip H. Howard* (blog), December 31, 2018, https://philhoward.net/2018/12/31/global-seed-industry-changes-since-2013/.

31. Author interview, Bill and Melinda Gates Foundation, Seattle, WA, July 23, 2015.

32. Author interview, Bill and Melinda Gates Foundation, Des Moines, IA, October 18, 2017.

33. Samuel A. Chambers, *There's No Such Thing as the Economy: Essays on Capitalist Value* (Santa Barbara, CA: Punctum Books, 2018), 41–57; Marx, *Capital*.

34. Charles W. Mills, "Global White Ignorance," in *Routledge International Handbook of Ignorance Studies* (London: Routledge, 2015).

35. Mills, 223.

36. James McCann, *Maize and Grace* (Cambridge, MA: Harvard University Press, 2005), 147–54; Matthew A. Schnurr, *Africa's Gene Revolution*, 110–13.

37. Nicole Marie Aschoff, *The New Prophets of Capital* (London: Verso, 2015).

38. Roy, "Introduction: the Aporias of Poverty."

39. For popular critiques, see Bill McKibben, "How Does Bill Gates Plan to Solve the Climate Crisis?" *New York Times*, February 15, 2021; Naomi Klein, *This Changes Everything: Capitalism vs. the Climate* (New York: Simon & Schuster, 2014). Tim Schwab, *The Bill Gates Problem: Reckoning with the Myth of the Good Billionaire* (New York: Metropolitan Books, 2023). For scholarly critique, see especially McGoey, *No Such Thing as a Free Gift*.

40. Amie Newman, "Gates Foundation Visitor Center: An Interview with Therese Littleton," Impatient Optimists, February 4, 2012, https://www.impatientoptimists.org/Posts/2012/02/Use-Your-Voice-for-Good-An-Interview-with-Therese-Littleton.

41. Kaiama L. Glover, "'Flesh Like One's Own': Benign Denials of Legitimate Complaint," *Public Culture* 29, no. 2 (May 1, 2017): 246, https://doi.org/10.1215/08992363-3749045.

42. Glover, 236.

43. Catherine Lutz and Jane Lou Collins, *Reading National Geographic* (Chicago: University of Chicago Press, 1993), 157.

44. Ananya Roy, *Poverty Capital: Microfinance and the Making of Development* (New York: Routledge, 2010).

45. Roopali Mukherjee, "Antiracism Limited: A Pre-History of Post-Race," *Cultural Studies* 30, no. 1 (January 2, 2016): 47–77, https://doi.org/10.1080/09502386.2014.935455.

46. Keith P. Feldman, "The Globality of Whiteness in Post-Racial Visual Culture," *Cultural Studies* 30, no. 2 (March 3, 2016): 289–311, https://doi.org/10.1080/09502386.2015.1020957.

47. Personal communication, Bill and Melinda Gates Foundation Discovery Center, March 20, 2018.

48. Crucial scholarship on postrace includes: Roopali Mukherjee, Sarah Banet-Weiser, and Herman Gray, eds., *Racism Postrace* (Durham, NC: Duke University Press, 2019); and Catherine R. Squires, *The Post-Racial Mystique: Media and Race in the Twenty-First Century* (New York: New York University Press, 2014).

49. Mark Landler, "Curing the Ills of America's Top Foreign Aid Agency," *New York Times*, October 23, 2010, sec. World, https://www.nytimes.com/2010/10/23/world/23shah.html.

50. McKibben, "How Does Bill Gates Plan to Solve the Climate Crisis?"

51. Laura Pulido, "Racism and the Anthropocene," in *Future Remains: A Cabinet of Curiosities for the Anthropocene*, ed. Gregg Mitman, Marco Armiero, and Robert S. Emmett (Chicago: University of Chicago Press, 2018), 116–28.

52. Glover, "'Flesh Like One's Own,'" 256.

CHAPTER 3. "THE LANDRACES ARE IN THE HYBRIDS"

1. For the project's first ten years, WEMA operated in Kenya, Uganda, Mozambique, Tanzania, and South Africa. With the Bayer acquisition of Monsanto in 2018, the project continued under the name TELA (to align with the branded hybrid and biotech seeds the project had developed). A third five-year funding phase for the renamed project began in 2018, with a $24.6 million grant from the Gates Foundation (to add to USAID's $5 million behind the project). See: AATF, "Press Release: AATF Receives Grant to Make New Drought-Tolerant and Insect-Resistant Maize Hybrids Available to Farmers in Africa," June 18, 2018, https://www.aatf-africa.org/wp-content/uploads/2018/11/Press-release-Gates-Foundation-Grants-AATF-24m.pdf.

2. Author interview, CIMMYT, Des Moines, Iowa, October 15, 2016.

3. Cynthia Hewitt de Alcantara, *Modernization of Mexican Agriculture* (Geneva: UNRISD, 1976); Joseph Cotter, *Troubled Harvest: Agronomy and Revolution in Mexico, 1880–2002* (Westport, CT: Praeger, 2003).

4. Nick Cullather, *The Hungry World: America's Cold War Battle against Poverty in Asia* (Cambridge, MA: Harvard University Press, 2013), 56. This is a simplified account of the beginning of the MAP, which has been widely covered in scholarship on the Green Revolution. For more comprehensive discussions, see especially chapter 2 in Cullather; Cotter, *Troubled Harvest*; Raj Patel, "The Long Green Revolution," *Journal of Peasant Studies* 40, no. 1 (January 2013):

1–63, https://doi.org/10.1080/03066150.2012.719224; and Tore C. Olsson, *Agrarian Crossings: Reformers and the Remaking of the US and Mexican Countryside* (Princeton, NJ: Princeton University Press, 2017).

5. The Commission noted that the Suburban was "originally red in color but was repainted green, possibly more in keeping with the mission." E. C. Stakman, Richard Bradfield, and Paul C Mangelsdorf, *Campaigns against Hunger* (Cambridge, MA: Belknap Press, 1967), 25.

6. Stakman et al., 31.

7. Stakman et al., 32.

8. Stakman et al., 2.

9. Stakman et al., 27.

10. *Agricultural Conditions and Problems in Mexico: Report of the Survey Commission of the Rockefeller Foundation*, Rockefeller Foundation, 1941, folder 37, box 5, series 323, R.G 1.1 Projects. Rockefeller Foundation Records, Rockefeller Archive Center, Tarrytown, New York. Subsequent citations are referenced in text by referring to the "commission" or their report.

11. Colin R. Johnson, *Just Queer Folks: Gender and Sexuality in Rural America* (Philadelphia: Temple University Press, 2013); Barbara A. Kimmelman, "The American Breeders' Association: Genetics and Eugenics in an Agricultural Context, 1903–13," *Social Studies of Science* 13, no. 2 (May 1, 1983): 163–204, https://doi.org/10.1177/030631283013002001.

12. Gabriel N. Rosenberg, "No Scrubs: Livestock Breeding, Eugenics, and the State in the Early Twentieth-Century United States," *Journal of American History* 107, no. 2 (September 1, 2020): 362–87, https://doi.org/10.1093/jahist/jaaa179.

13. María Josefina Saldaña-Portillo, *Indian Given: Racial Geographies across Mexico and the United States* (Durham, NC: Duke University Press Books, 2016).

14. John H. Perkins, *Geopolitics and the Green Revolution: Wheat, Genes, and the Cold War* (New York: Oxford University Press, 1997), 104.

15. Confidential Report: *Rockefeller Foundation Scholarships and the Mexican Revolution in Agricultural Science*," Rockefeller Foundation, 1959, folder 22, box 5, E. C. Stakman Papers, University of Minnesota Archives, Minneapolis, Minnesota.

16. *Report of E. C. Stakman, Trip to Columbia, Ecuador, Peru, and Mexico, June 30–August 6, 1953*, Rockefeller Foundation, 1953, Folder 31, box 5, E. C. Stakman Papers, University of Minnesota Archives, Minneapolis, Minnesota.

17. E. C. Stakman, "Report on Mexico, Jan. 8–April 9, 1960: The Agricultural Revolution." E. C. Stakman Papers, box 4, folder 12. University of Minnesota Archives, Minneapolis, MN.

18. Kenneth Wernimont, "Confidential Report: *Rockefeller Foundation Scholarships and the Mexican Revolution in Agricultural Science*," no. 197, April 1959, University of Minnesota Archives, E. C. Stakman Papers, box 5, folder 22.

19. Robert Lee and Tristan Ahtone, "Land-Grab Universities," *High Country News*, March 30, 2020, https://www.hcn.org/issues/52.4/indigenous-affairs-education-land-grab-universities.

20. Roderick A. Ferguson, *The Reorder of Things: The University and Its Pedagogies of Minority Difference* (Minneapolis: University of Minnesota Press, 2012).

21. Bruce H. Jennings, *Foundations of International Agricultural Research: Science and Politics in Mexican Agriculture* (Boulder, CO: Westview Press, 1988), 104.

22. Stakman et al., *Campaigns against Hunger*, 40.

23. Wellhausen's Rockefeller Foundation Officer Diaries detail extensive correspondence with company leadership from the burgeoning American hybrid seed industry. Wellhausen sent maize varieties to companies like Pioneer, DeKalb, and Northrop King.

24. Stakman et al., *Campaigns against Hunger*, 58.

25. "Edwin J. Wellhausen Oral History," Rockefeller Foundation, June 28, 1966, Folder 1, Box 25, RG 13, Oral Histories, FA119, Rockefeller Foundation records, Rockefeller Archive Center, Tarrytown, New York, 141. Subsequent citations are in text by page number.

26. Saldaña-Portillo, *Indian Given*, 9.

27. Hernández Xolocotzi was later critical of the MAPs earlier work and worked to develop maize collection projects that considered indigenous knowledge and culture. See Helen Anne Curry, "Taxonomy, Race Science, and Mexican Maize," *Isis* 112, no. 1 (March 1, 2021): 1–21, https://doi.org/10.1086/713819.

28. Reflecting their keen interested in genetic heritage, they also doubted that he was "pure-bred" Aztec, as he reportedly claimed, because he had "blue eyes." See "Neil B. Manglesdorf Oral History," Rockefeller Foundation, November 16 and December 19, 1966, Folder 1, Box 18, RG 13, Oral Histories, FA119, Rockefeller Foundation Records, Rockefeller Archive Center, Tarrytown, New York, 65.

29. Jean M. O'Brien, *Firsting and Lasting: Writing Indians Out of Existence in New England* (Minneapolis: University of Minnesota Press, 2010).

30. Donna Haraway reminds us that "a gene is not a thing" in itself, but must be made legible through social and technical processes. This does not mean that they are not real, she insists. Being "made" is not the same as being "made up." "A gene is not a thing," she writes, "much less a 'master molecule' or a self-contained code. Instead, the term *gene* signifies a node of durable action where many actors, human and nonhuman, meet." Donna Haraway, *Modest_Witness@Second_Millennium.FemaleMan_Meets_OncoMouse: Feminism and Technoscience* (New York: Routledge, 1997), 142.

31. E. J. Wellhausen, "Exotic Germ Plasm for Improvement of Corn Belt Maize," in *Proceedings of the 20th Annual Hybrid Corn Industry-Research Conference* (Chicago, 1965), 43. Subsequent quotations from Wellhausen's talk are cited in text.

32. On the logic of elimination, see Patrick Wolfe, "Settler Colonialism and the Elimination of the Native," *Journal of Genocide Research* 8, no. 4 (December 2006): 387–409, https://doi.org/10.1080/14623520601056240. In addition to the scholars cited below, Sisseton Wahpeton Oyate scholar Kim TallBear's work

is especially generative for thinking about science, race, and settler colonialism. See Kim TallBear, *Native American DNA: Tribal Belonging and the False Promise of Genetic Science* (Minneapolis: University of Minnesota Press, 2013).

33. Maile Arvin, *Possessing Polynesians: The Science of Settler Colonial Whiteness in Hawai'i and Oceania* (Durham, NC: Duke University Press, 2019), 24.

34. Aileen Moreton-Robinson, *The White Possessive: Property, Power, and Indigenous Sovereignty* (Minneapolis: University of Minnesota Press, 2015), 126.

35. Moreton-Robinson, 114.

36. E. J. Wellhausen, *Races of Maize in Mexico: Their Origin, Characteristics and Distribution* (Cambridge, MA: Bussey Institution of Harvard University, 1952).

37. Wellhausen,. 44.

38. For an explanation of *co*-production, see: Sheila Jasanoff, ed., *States of Knowledge: The Co-production of Science and Social Order* (London: Routledge, 2010). I utilized Jasanoff's framework in a previous iteration of this chapter: Aaron Eddens, "White Science and Indigenous Maize: The Racial Logics of the Green Revolution," *Journal of Peasant Studies* 46, no. 3 (April 16, 2019): 653–73, https://doi.org/10.1080/03066150.2017.1395857. For another useful engagement of the concept, see: Rosenberg, "No Scrubs."

39. Olsson, *Agrarian Crossings*, 154.

40. *Director's Annual Report: September 1958–August 1959: Mexican Agriculture Program* (New York: The Rockefeller Foundation, 1959), 23–42. https://www.rockefellerfoundation.org/wp-content/uploads/Annual-Report-1959.pdf.

41. Deborah Fitzgerald, "Exporting American Agriculture: The Rockefeller Foundation in Mexico, 1943–53," *Social Studies of Science* 16, no. 3 (1986): 467.

42. "J. G. Harrar Oral History," Rockefeller Foundation, 1961, Folder 308, Box 42, Subseries 2.2, series 2, FA046, Rockefeller Foundation records, Rockefeller Archive Center, Tarrytown, New York, 188.

43. E. J. Wellhausen to Loring Jones, June 9, 1961, Folder 162, Box 14, Series 1.1, RG 6.13, Rockefeller Foundation records, RAC.

44. "Elmer Johnson Oral History," Rockefeller Foundation, Box 17, RG 13, Oral histories, Rockefeller Foundation records, RAC, 38–39 and 59–60.

45. Pioneer, "History of Pioneer," https://www.pioneer.com/us/about-us/our-history.html.

46. Derek Byerlee, "The Globalization of Hybrid Maize, 1921–70," *Journal of Global History* 15, no. 1 (March 2020): 105, https://doi.org/10.1017/S1740022819000354.

47. Carl E. Pray and Ruben G. Echeverria, "Transferring Hybrid Maize Technology: The Role of the Private Sector," *Food Policy* 13, no. 4 (November 1, 1988): 366–74, https://doi.org/10.1016/0306-9192(88)90084-X.

48. Jack R. Kloppenburg Jr., *First the Seed: The Political Ecology of Plant Biotechnology, 1492–2000*. 2nd ed. (Madison: University of Wisconsin Press, 2004), chap. 7.

49. For a useful discussion of how White anthropologists naturalized ideas about appropriating indigenous peoples' land and material culture in the United States, see Yael Ben-zvi, "Where Did Red Go? Lewis Henry Morgan's Evolutionary Inheritance and U.S. Racial Imagination," *CR: The New Centennial Review* 7, no. 2 (2007): 201–29, https://doi.org/10.1353/ncr.2007.0037.

50. For a discussion of the politics of hybrid and GM maize in Mexico, see Elizabeth M. Fitting, *The Struggle for Maize: Campesinos, Workers, and Transgenic Corn in the Mexican Countryside* (Durham, NC: Duke University Press, 2011).

51. Enrique Perez, "Collective Action, Democracy and Mexico's Defense of Its Corn and Food Sovereignty," Institute for Agriculture and Trade Policy, December 20, 2022, https://www.iatp.org/collective-action-democracy-mexicos-defense-corn-food-sovereignty.

52. A recent history of maize projects through CIMMYT details eleven projects across Africa since 2007. Derek Byerlee and Greg Edmeades, *Fifty Years of Maize Research in the CGIAR* (Mexico City: CIMMYT, 2021).

53. George Mahuku et al., "Maize Lethal Necrosis (MLN), an Emerging Threat to Maize-Based Food Security in Sub-Saharan Africa," *Phytopathology* 105, no. 7 (March 30, 2015): 956–65, https://doi.org/10.1094/PHYTO-12-14-0367-FI.

54. CIMMYT, "New Screening Cycle Begins for Maize Lethal Necrosis Disease in Kenya," October 3, 2017, https://www.cimmyt.org/news/new-screening-cycle-for-deadly-mln-virus-set-to-begin-in-kenya/.

55. Author interview, CIMMYT, Des Moines, Iowa, October 13, 2016.

56. Maywa Montenegro, "Opinion: CRISPR Is Coming to Agriculture—with Big Implications for Food, Farmers, Consumers and Nature," *Ensia*, January 28, 2016, https://ensia.com/voices/crispr-is-coming-to-agriculture-with-big-implications-for-food-farmers-consumers-and-nature/.

57. The CRISPR patent story is somewhat complicated. A few institutions hold initial patents, and they license them to a number of companies. Currently, the ag-biotech firm Corteva is the largest holder of patents on applications of CRISPR. Julie Deering, "Who Owns CRISPR?," *Germination*, November 16, 2018, https://germination.ca/who-owns-crispr/.

58. CIMMYT, "DuPont Pioneer and CIMMYT Form CRISPR-Cas Public/Private Partnership," CIMMYT, September 28, 2016, http://www.cimmyt.org/press_release/dupont-pioneer-and-cimmyt-form-crispr-cas-publicprivate-partnership/Sept. 28.

59. "CIMMYT, CRISPR-CAS Technology by Neal Guterson [sic]," 2016, https://www.youtube.com/watch?v=goOmHFQyoXI&list=PLjXdzeDP_y5FEhl8XKm7dosQbArng7mn3&index=21.

60. Maywa Montenegro de Wit, "Stealing into the Wild: Conservation Science, Plant Breeding and the Makings of New Seed Enclosures," *Journal of*

Peasant Studies 44, no. 1 (January 2, 2017): 169–212, https://doi.org/10.1080
/03066150.2016.1168405.

61. T. Garrett Graddy, "Situating in Situ: A Critical Geography of Agricultural Biodiversity Conservation in the Peruvian Andes and Beyond," *Antipode* 46, no. 2 (March 2014): 426–54, https://doi.org/10.1111/anti.12045.

CHAPTER 4. SEEING LIKE A SEED COMPANY

1. Author interview, Monsanto Company, St. Louis, MO, May 26, 2015.
2. Emily Waltz, "Beating the Heat," *Nature Biotechnology* 32, no. 7 (July 2014): 610–13, https://doi.org/10.1038/nbt.2948.
3. The Howard Buffett Foundation was also an original funder but is not currently funding the project. USAID began funding the project during its second phase, which began in 2013.
4. African Agricultural Technology Foundation, "Concept Note: Water Efficient Maize for Africa," accessed June 19, 2015, http://www.aatf-africa.org.
5. The *Bt* trait was added when the project began its second phase. My arguments about the gene-as-property apply equally to this biotech trait, but because the framing of the project so often focuses on the drought trait, I primarily focus on that technology.
6. Katherine Verdery, *The Vanishing Hectare: Property and Value in Postsocialist Transylvania* (Ithaca, NY: Cornell University Press, 2003).
7. William Boyd, "Wonderful Potencies? Deep Structure and the Problem of Monopoly in Agricultural Biotechnology," in *Engineering Trouble: Biotechnology and Its Discontents*, ed. Rachel A. Schurman and Dennis Doyle Takahashi Kelso (Berkeley: University of California Press, 2003), 24–62.
8. These included representatives from Monsanto, AATF, CIMMYT, Bill and Melinda Gates Foundation, USAID, Kenya Agricultural Research Institute, and three Kenyan seed companies.
9. Author interview, Monsanto Company, St. Louis, MO, May 26, 2015.
10. Author interview, Monsanto Company, St. Louis, MO, May 26, 2015.
11. Author interview, Monsanto Company, St. Louis, MO, June 26, 2015.
12. Author interview, AATF, Nairobi, Kenya, August 26, 2015.
13. Author interview, Monsanto Company, St. Louis, MO, May 26, 2015.
14. Author interview, Bill and Melinda Gates Foundation, Seattle, WA, July 22, 2015.
15. Author interview, AATF, Nairobi, Kenya, August 17, 2015.
16. Author interview, CIMMYT, Nairobi, Kenya, September 2, 2015.
17. For example, Nick Cullather points to the US firms that directly benefitted from the GR development programs. Nick Cullather, "Miracles of Modernization: The Green Revolution and the Apotheosis of Technology," *Diplomatic*

History 28, no. 2 (2004): 227–54. As Jack Kloppenburg argues, the expansion of hybrid seeds is also the expansion of private property in seed systems. Jack R. Kloppenburg, *First the Seed: The Political Ecology of Plant Biotechnology, 1492–2000*. 2nd ed. (Madison: University of Wisconsin Press, 2004).

18. "Confidential Report: Rockefeller Foundation Scholarships and the Mexican Revolution in Agricultural Science," Rockefeller Foundation, 1959, E.C. folder 22, box 5, Box 5, Stakman Papers, University of Minnesota Archives, Minneapolis, Minnesota, emphasis added.

19. "Report on Mexico, January 8–April 9, 1960: The Agricultural Revolution," Rockefeller Foundation, 1960, folder 12, box 4, E. C. Stakman Papers, University of Minnesota Archives, Minneapolis, Minnesota.

20. Cullather, "Miracles of Modernization, 237.

21. Chris J. Shepherd, "Imperial Science: The Rockefeller Foundation and Agricultural Science in Peru, 1940–1960," *Science as Culture* 14, no. 2 (2005): 119–20.

22. Raj Patel, "The Long Green Revolution," *Journal of Peasant Studies* 40, no. 1 (January 2013): 1–63, https://doi.org/10.1080/03066150.2012.719224.

23. Ellen Meiksins Wood, "The Agrarian Origins of Capitalism," in *Hungry for Profit*, ed. Frederick H. Buttel, Fred Magdoff, and John Bellamy Foster (New York: Monthly Review Press, 2000), 23–41; John Locke, *The Second Treatise of Government and A Letter Concerning Toleration* (Mineola, NY: Dover, 2002), chap. 5, sect. 34.

24. Peter Linebaugh and Marcus Rediker, *The Many-Headed Hydra: Sailors, Slaves, Commoners, and the Hidden History of the Revolutionary Atlantic*, reprint ed. (Boston: Beacon Press, 2013), chap. 1.

25. Brenna Bhandar, *Colonial Lives of Property: Law, Land, and Racial Regimes of Ownership* (Durham, NC: Duke University Press, 2018), 36.

26. Bhandar, 9.

27. I developed this link between today's Green Revolution and the history of the improvement logic in my 2019 dissertation. More recently, Matthew Canfield has made similar arguments to connect the ideology of improvement to the Gates Foundation's "ideology of innovation." See: Matthew Canfield, "The Ideology of Innovation: Philanthropy and Racial Capitalism in Global Food Governance," *Journal of Peasant Studies*, September 19, 2022, 1–25, https://doi.org/10.1080/03066150.2022.2099739.

28. This chapter's title is a riff on James Scott's classic work on how "development" projects rooted in a "high modern" perspective discount local knowledge—and often fail to "improve the human condition" in the process. James C. Scott, *Seeing Like a State: How Certain Schemes to Improve the Human Condition Have Failed* (New Haven, CT: Yale University Press, 2008).

29. Silvia Federici, *Re-enchanting the World: Feminism and the Politics of the Commons* (Oakland, CA: PM Press, 2019), 87.

30. Egypt and Burkina Faso would soon deregulate varieties of Monsanto's GM cotton. Egypt in 2008 and Burkina Faso in 2009. See ISAAA GM approval database: http://www.isaaa.org/gmapprovaldatabase/.

31. Rachel Schurman and William A. Munro, *Fighting for the Future of Food: Activists versus Agribusiness in the Struggle over Biotechnology* (Minneapolis: University of Minnesota Press, 2010).

32. Schurman and Munro.

33. Author interview, Bill and Melinda Gates Foundation, Seattle, WA, July 21, 2015.

34. Don S. Doering, "Public-Private Partnership to Develop and Deliver Drought Tolerant Crops to Food-Insecure Farmers." Summary and Interpretation of the May 3–4 Strategy and Planning Meeting, Winrock International, May 31, 2005, 2.

35. Doering, 1.

36. Doering, 5.

37. Doering, 4.

38. Throughout Doering's summary document, "poor" and "rural" are frequently used to describe smallholder farmers. I include the terms here to demonstrate how these farmers are consistently figured in terms that emphasize what they lack.

39. Doering, "Public-Private Partnership," 4.

40. Doering, 4.

41. Author interview, Bill and Melinda Gates Foundation, Seattle, WA, July 22, 2015.

42. Author interview, Gates Foundation, July 22, 2015.

43. Author interview, Gates Foundation, July 22, 2015.

44. A longtime veteran of international agricultural development projects, Robert Herdt, who himself worked at the Gates, Ford, and Rockefeller Foundation, wrote in 2012 that the Gates Foundation's agricultural development program was larger than those of the World Bank and USAID. Robert W. Herdt, "People, Institutions, and Technology: A Personal View of the Role of Foundations in International Agricultural Research and Development 1960–2010," *Food Policy* 37, no. 2 (April 2012): 179–90, https://doi.org/10.1016/j.foodpol.2012.01.003.

45. The Foundation engaged several officials from Monsanto during the planning of its agriculture program. As a Gates Foundation official explained to me, as the foundation was considering an agriculture program, leadership in the Foundation's global development group had several generative conversations with a retired Monsanto CEO. The Gates Foundation sought the expertise of Monsanto leadership in their earliest discussion about what their agricultural development program might look like. Author interview, Bill and Melinda Gates Foundation, Seattle, WA, July 21, 2015.

46. Horsch, "Reflections of a Science Pioneer," *Monsanto News*, January 31, 2006.

47. Because of his association with Monsanto, Horsch does not work on the WEMA project; Monsanto/AATF, "Combining Breeding and Biotechnology to Develop Drought Tolerant Maize for Africa: A Proposal to the Bill and Melinda Gates Foundation," May 25, 2007, St. Louis, MO.

48. Robert L. Paarlberg, *Starved for Science: How Biotechnology Is Being Kept Out of Africa* (Cambridge, MA: Harvard University Press, 2009), 170.

49. A Monsanto representative explained that the company began a drought-tolerant maize development program in South Africa in 2005. Under the name "Project Rain Barrel," they tested hybrid varieties of maize developed in the United States at several locations across the country. They planted their first trials of GM crops in 2007. Prior to the WEMA partnerships launch in 2008, most of the work on hybrid and GM maize in South Africa was done by Monsanto employees. Personal communication, Monsanto Company, July 1, 2015.

50. Author interview, Monsanto Company, St. Louis, MO, May 26, 2015.

51. Author interview, Bill and Melinda Gates Foundation, Seattle, WA, July 22, 2015.

52. Author interview, AATF, Des Moines IA, October 14, 2014.

53. The network is known as the CGIAR. As part of its recent organizational reforms, it officially dropped the title of "Consultative Group for International Agricultural Research" and became known as simply the CGIAR.

54. Schurman and Munro, *Fighting for the Future of Food*, 40–50.

55. Author interview, Bill and Melinda Gates Foundation, Seattle, WA, July 21, 2015.

56. As one of my interviewees shared, the jingle was in the tune of "The Lion Sleeps Tonight," made famous in the United States by the Tokens in 1961. Apparently, an astute team-building facilitator turned the song's memorable opening lyrics (A-weema-weh, a-weema-weh, a-weema-weh, a-weema-weh) into "A WEMA way."

57. Author interview, Bill and Melinda Gates Foundation, Seattle, WA, July 22, 2015.

58. Author interview, Gates Foundation, Seattle.

59. Author interview, AATF, Des Moines, IA, October 14, 2014; Author interview, Monsanto Company, St. Louis, MO, June 26, 2015.

60. Rachel Schurman, "Building an Alliance for Biotechnology in Africa," *Journal of Agrarian Change* 17, no. 3 (July 2017): 441–58, https://doi.org/10.1111/joac.12167.

61. Schurman, 12.

62. Author interview, Monsanto Company, St. Louis, MO, May 26, 2015; Author interview, CIMMYT, July 31, 2015.

63. Author interview, Monsanto East Africa, Nairobi, Kenya, August 20, 2015. Here is how a WEMA official described how the drought gene was licensed in

the project: "AATF negotiated with Monsanto for the donation of the [drought trait] into the WEMA project and then Monsanto accepted and offered the technology to WEMA as a project. Because this is a public-private partnership meant to benefit smallholder farmers, the royalty aspect has been eliminated by the virtue of the AATF [negotiating] access to proprietary technologies that otherwise smallholder farmers would not have accessed because of the requirements to pay royalties, . . . and when the technology arrives to AATF it brings these [seed companies] on board, they pick up the technology, and they develop a product out of it, and that product is not subjected to royalties because AATF did the negotiation and advocacy from the technology owner [Monsanto]."

64. Author interview, Monsanto Company, St. Louis, MO, May 26, 2015.

65. Author interview, AATF, Des Moines IA, October 14, 2014. As company lawyers explained much of Monsanto's business is built around "trade secrets"—information that they cannot really legally protect with patents, but that they are careful not to share.

66. Author interview, AATF; Author interview, Bill and Melinda Gates Foundation, Seattle, WA, July 22, 2015.

67. Author interview, AATF, Nairobi, Kenya, August 10, 2015.

68. Author interview, AATF, Nairobi.

69. Author interview, CIMMYT, July 31, 2015.

70. African Agricultural Technology Foundation, n.d., "Concept Note: Water Efficient Maize for Africa Program."

71. Author interview, AATF, Nairobi, Kenya, August 17, 2015. This last point, in which my interviewee staged a possible conversation with a politician, would have, at the time, applied to a country like Uganda, where WEMA officials were advocating for the passage of a biosafety policy. Kenya already had a biosafety policy in effect at the time of this interview (2015) but had not deregulated any biotech crops. WEMA has the capacity to push regulatory systems in different ways, whether advocating for passing laws or pushing the system to approve products.

72. Author interview, Monsanto Company, St. Louis, MO, June 26, 2015.

73. Much of the media coverage of the finalizing of Bayer's acquisition of Monsanto played up the "end" of the much-maligned company, as this NPR headline suggests: Camila Domonoske, "Monsanto No More: Agri-Chemical Giant's Name Dropped in Bayer Acquisition," *The Two-Way*, NPR, June 4, 2018, https://www.npr.org/sections/thetwo-way/2018/06/04/616772911/monsanto-no-more-agri-chemical-giants-name-dropped-in-bayer-acquisition.

74. Ruth Bender, "How Bayer-Monsanto Became One of the Worst Corporate Deals—in 12 Charts," *Wall Street Journal*, August 28, 2019, sec. Business, https://www.proquest.com/docview/2281135814/citation/3988E25F879E49D8PQ/3.

75. Ludwig Burger and Patricia Weiss, "Bayer's Agriculture Unit, Consumer Health Drive Outlook Hike," *Reuters*, August 4, 2022, https://www.reuters

.com/business/healthcare-pharmaceuticals/bayers-agriculture-unit-consumer-health-drives-improved-outlook-2022-2022-08-04/.

76. Winnie Nanteza, "WEMA Achieves Major Milestone in African Agriculture," Alliance for Science, May 29, 2018, https://allianceforscience.org/blog/2018/05/wema-achieves-major-milestone-african-agriculture/.

77. Quoted in Nanteza.

78. AATF, "TELA Maize Technology: FAQs," February 2021, https://www.aatf-africa.org/wp-content/uploads/2021/02/TELA-Project-FAQ.pdf.

79. Author interview, Monsanto Company, St. Louis, MO, 5-26-15.

80. Jaco Visser, "Monsanto Targets Smallholder Farmers," *Farmer's Weekly*, January 8, 2015, https://www.farmersweekly.co.za/bottomline/monsanto-targets-smallholder-farmers/.

81. Author interview, AATF, Nairobi, Kenya, August 17, 2015.

82. Author interview, Monsanto Company, St. Louis, MO, May 27, 2015.

83. Author interview, Monsanto East Africa, Nairobi, Kenya, August 20, 2015.

84. One of WEMA officials' biggest concerns is whether smallholder farmers will be able to adhere to the requirement to plant a "refuge" area when planting *Bt* GM crops. Companies like Monsanto require farmers to plant a certain percentage of their fields in non-*Bt* crops, to lessen insects' ability to develop resistance to *Bt*'s poisonous effect.

85. Author interview, CIMMYT, Nairobi, Kenya, September 2, 2015.

86. Doering, "Public-Private Partnership," 4.

CHAPTER 5. SECURITIZING SMALLHOLDER FARMERS ON THE FRONT LINES OF THE CLIMATE CRISIS

1. Jane Bird, "'Smart' Insurance Helps Poor Farmers to Cut Risk," *Financial Times*, December 5, 2018, https://www.ft.com/content/3a8c7746-d886-11e8-aa22-36538487e3d0; "In Africa, Agricultural Insurance Often Falls on Stony Ground," *Economist*, December 15, 2018, https://www.economist.com/finance-and-economics/2018/12/15/in-africa-agricultural-insurance-often-falls-on-stony-ground; Norman Mayersohn, "He Grew Up on a Farm; Now, He Helps Protect Them," *New York Times*, October 3, 2019. https://www.nytimes.com/2019/10/03/business/microinsurance-africa-thomas-njeru.html.

2. Pula, "Global Insuretech," accessed September 30, 2021, https://www.pula-advisors.com.

3. As I discussed in chapter 4, these companies, now the two largest agricultural-biotechnology/agri-chemical companies in the world resulted from two high-profile mergers. Dow and DuPont merged in 2017; Bayer acquired Monsanto in 2018.

4. Daniela Gabor and Sally Brooks, "The Digital Revolution in Financial Inclusion: International Development in the Fintech Era," *New Political Economy* 22, no. 4 (July 4, 2017): 424, https://doi.org/10.1080/13563467.2017.1259298.

5. I use the term *security state* not just to refer to the parts of the state most commonly associated with "national security" or "defense," but to signal more broadly the increasing *orientation* toward security thinking that has become pervasive across state institutions. Inderpal Grewal, *Saving the Security State: Exceptional Citizens in Twenty-First-Century America* (Durham, NC: Duke University Press, 2017).

6. National Intelligence Council, "Intelligence Community Assessment: Global Food Security," September 2015, i. https://www.dni.gov/files/documents/Newsroom/Reports%20and%20Pubs/Global_Food_Security_ICA.pdf. An NIC official that spoke at the 2016 World Food Prize Conference in Des Moines, Iowa, mentioned that these NIC assessments were unclassified derivatives of classified reports. Author field notes.

7. National Intelligence Council, "Implications for US National Security of Anticipated Climate Change," September 21, 2016, 8–9, https://www.dni.gov/files/documents/Newsroom/Reports%20and%20Pubs/Implications_for_US_National_Security_of_Anticipated_Climate_Change.pdf.

8. National Intelligence Council, *Global Trends: A Paradox of Progress*, National Intelligence Council, Office of the Director of National Intelligence, 2017, https://www.cisa.gov/sites/defaultfiles/publications/global-trends-paradox%20-of-progress-508.pdf.

9. Suzanne Fry, "National Security Is Food Security: Strategic Leadership and a Moral Imperative, Panel Remarks," Global Food Security Symposium, Chicago Council on Global Affairs, Washington, D.C., March 30, 2017.

10. Jeremy Walker and Melinda Cooper, "Genealogies of Resilience: From Systems Ecology to the Political Economy of Crisis Adaptation," *Security Dialogue* 42, no. 2 (2011): 143–60. https: doi.org/10.1177/0967010611399616.

11. Walker and Cooper, 144.

12. National Intelligence Council, *Global Trends*, xi.

13. Walker and Cooper, "Genealogies of Resilience," 156.

14. Grewal, *Saving the Security State*.

15. Grewal, 16–17.

16. U.S. Agency for International Development, "Why Resilience?," video, accessed September 30, 2021, https://www.usaid.gov/resilience.

17. H.R 1567—Global Food Security Act of 2016. H.R. 1567, 114th Congress, December 4, 2016, https://www.congress.gov/bill/114th-congress/house-bill/1567.

18. Jamey Essex, *Development, Security, and Aid: Geopolitics and Geoeconomics at the U.S. Agency for International Development* (Athens: University of Georgia Press, 2013), 147.

19. Speaking at a symposium at the Center for Strategic and International Studies in 2016, Adele Adeyemo, the deputy assistant to the president and deputy national security adviser for international economics, stated that the national security framing of the Global Food Security Act had been crucial for getting it

passed with sweeping bipartisan support. See: Center for Strategic and International Studies, "The Power of Global Food Security: Examining Economic and National Security Implications," CSIS Presents, September 12, 2016, https:/www.csis.org/events/power-global-food-security-examining-economic-and-national-security-implications.

20. US Agency for International Development, "U.S. Government Global Food Security Strategy: FY 2017–2021," September 2016, iii, https://www.usaid.gov/sites/default/files/Global-Food-Security-Strategy_2017-2021.pdf.

21. USAID, 18.

22. Walker and Cooper. "Genealogies of Resilience," 154.

23. USAID, "U.S. Government Global Food Security Strategy."

24. USAID, 24.

25. USAID, 30.

26. USAID, 18.

27. Author interview, Syngenta Foundation, 2015.

28. Author interview, Monsanto Company, St. Louis, MO, May 26, 2015.

29. "De-Risking Agricultural Investment in Africa," *Financial Times*, August 14, 2018, https:/www.ft.com/content/4ee682ec-9fd6-11e8-85da-eeb7a9ce36e4.

30. Pula's website describes this process: "We increase credit providers' appetite for lending to farmers by offering a safety net." "About Us." n.d. Pula. Accessed October 1, 2021. https://www.pula-advisors.com/about.

31. Jonathan Levy, *Freaks of Fortune: The Emerging World of Capitalism and Risk in America* (Cambridge, MA: Harvard University Press, 2012).

32. Leigh Johnson, "Index Insurance and the Articulation of Risk-Bearing Subjects," *Environment & Planning A* 45, no. 11 (2013): 2663–81, https://doi.org/10.1068/a45695.

33. Author interview, Development Organization, January 20, 2015.

34. Johnson, "Index Insurance and the Articulation of Risk-Bearing Subjects," 2667.

35. USAID, "U.S. Government Global Food Security Strategy," 8.

36. Helen Greatrex et al. "Scaling Up Index Insurance for Smallholder Farmers: Recent Evidence and Insights," CCAFS Working Paper, 2015, https://hdl.handle.net/10568/53101.

37. Author interview, Syngenta Foundation, January 22, 2015.

38. Costas Lapavitsas, *Profiting without Producing: How Finance Exploits Us All* (New York: Verso, 2013), 5.

39. Melinda Cooper, "Turbulent Worlds: Financial Markets and Environmental Crisis," *Theory, Culture & Society* 27, nos. 2–3 (2010): 167–90, https://doi.org/10.1177/0263276409358727.

40. Johnson, "Index Insurance and the Articulation of Risk-Bearing Subjects," 2665.

41. S. Ryan Isakson, "Derivatives for Development? Small-Farmer Vulnerability and the Financialization of Climate Risk Management," *Journal of Agrarian Change* 15, no. 4 (2015): 569–80. https://doi.org/10.1111/joac.12124.

42. María Josefina Saldaña-Portillo argues that "racial geography is a technology of power, and when used as an analytic and theory of spatial production it indexes the series of techniques used to produce space in racial terms." María Josefina Saldaña-Portillo, *Indian Given: Racial Geographies across Mexico and the United States* (Durham, NC: Duke University Press, 2016), 17.

43. Saldaña-Portillo, 17 (emphasis in original).

44. Saldaña-Portillo, 18.

45. Clive Gabay, *Imagining Africa: Whiteness and the Western Gaze* (Cambridge: Cambridge University Press, 2018); Catherine Lutz, and Jane Lou Collins, *Reading National Geographic* (Chicago: University of Chicago Press, 1993).

46. Kaiama L. Glover, "'Flesh Like One's Own': Benign Denials of Legitimate Complaint," *Public Culture* 29, no. 2 (2017): 241.

47. Glover, 243.

48. Katherine McKittrick, "Plantation Futures," *Small Axe: A Caribbean Journal of Criticism* 17, no. 3 (2013): 7, https://doi.org/10.1215/07990537-2378892.

49. Glover, "'Flesh Like One's Own,'" 242.

50. National Intelligence Council, *Global Trends*, 119.

51. Gabor and Brooks, "The Digital Revolution in Financial Inclusion," 429.

52. Raj Patel and Jason W. Moore, *A History of the World in Seven Cheap Things: A Guide to Capitalism, Nature, and the Future of the Planet* (Oakland: University of California Press, 2017), 19.

53. Jodi Melamed, "Racial Capitalism," *Critical Ethnic Studies* 1, no. 1 (2015): 77, https:/doi.org/10.5749/jcritethnstud.1.1.0076.

54. Melamed, 77.

55. Paula Chakravartty and Denise Ferreira da Silva, "Accumulation, Dispossession, and Debt: The Racial Logic of Global Capitalism—An Introduction," *American Quarterly* 64, no. 3 (2012): 361–85, https://doi.org/10.1353/aq.2012.0033.

56. Ananya Roy, *Poverty Capital: Microfinance and the Making of Development* (New York: Routledge, 2010).

57. Roy, 218.

58. Chakravartty and Ferreira da Silva, "Accumulation, Dispossession, and Debt," 363.

59. Chakravartty and Ferreira da Silva, 364 (emphasis in original).

60. Chakravartty and Ferreira da Silva, 369.

61. Zenia Kish, and Justin Leroy, "Bonded Life: Technologies of Racial Finance from Slave Insurance to Philanthrocapital," *Cultural Studies* 29, nos. 5–6 (2015): 632, https://doi.org10.1080/09502386.2015.1017137.

62. Kish and Leroy, 633.

63. Kish and Leroy, 646.

64. Johnson, "Index Insurance and the Articulation of Risk-Bearing Subjects," 2663.

65. Isakson, "Derivatives for Development?"

66. Author interview, Syngenta Foundation, January 22, 2015.

67. Fry, "National Security Is Food Security."

68. Cooper, "Turbulent Worlds," 167–90.

69. Nick Turse, *Tomorrow's Battlefield: U.S. Proxy Wars and Secret Ops in Africa* (Chicago: Haymarket Books, 2015).

70. Turse, 3.

71. Nick Turse, "Exclusive: The U.S. Has More Military Operations in Africa than the Middle East," *Vice News*, December 12, 2018, https://www.vice.com/en_us/article/a3my38/exclusive-the-us-has-more-military-operations-in-africa-than-the-middle-east. The scholarship and public writing of anthropologist Samar Al-Bulushi is also crucial on the issue of military partnerships between the US and African nations. See, for example: Samar Al-Bulushi, "Geographies of war-making in East Africa," *Africa is a Country*, April 15, 2021. https://africasacountry.com/2021/04/geographies-of-war-making-in-east-africa.

72. Nick Turse, "The U.S. Military Moves Deeper into Africa," *TomDispatch*, April 27, 2017, http://www.tomdispatch.com/post/176272/tomgram%3A_nick_turse%2C_the_u.s._military_moves_deeper_into_africa.

73. Nick Turse, Sam Mednick, and Amanda Sperber, "Exclusive: Inside the Secret World of US Commandos in Africa," Pulitzer Center, August 11, 2020. https://pulitzercenter.org/reportingexclusive-inside-secret-world-us-commandos-africa.

74. John R. Bolton, "Remarks by National Security Advisor Ambassador John R. Bolton on the Trump Administration's New Africa Strategy," The White House, accessed September 14, 2020, https://www.whitehouse.gov/briefings-statements/remarks-national-security-advisor-ambassador-john-r-bolton-trump-administrations-new-africa-strategy.

75. Randy Martin, *An Empire of Indifference: American War and the Financial Logic of Risk Management* (Durham NC: Duke University Press, 2007), 18.

76. Walker and Cooper, "Genealogies of Resilience," 152.

CONCLUSION

1. "The Global Climate Wall: Wealthy Nations Prioritize Militarizing Borders over Climate Action," Democracy Now!, November 10, 2021, https://www.democracynow.org/2021/11/10/global_climate_wall.

2. Christopher Flavelle et al., "Climate Change Poses a Widening Threat to National Security," *New York Times*, October 21, 2021, sec. Climate, https://www.nytimes.com/2021/10/21/climate/climate-change-national-security.html.

3. Raj Patel, "The Long Green Revolution," *Journal of Peasant Studies* 40, no. 1 (January 2013): 1–63, https://doi.org/10.1080/03066150.2012.719224.

4. Author interview, Monsanto Company, St. Louis, MO, May 27, 2015.

5. Kelly Bronson, "The Dangers of Big Data Extend to Farming," The Conversation, June 27, 2022, http://theconversation.com/the-dangers-of-big-data-extend-to-farming-184531.

6. "LIVE: CNBC's Steve Sedgwick Joins Davos Panel on Revolutionizing Food Security—1/18/23," video, 2023, https://www.youtube.com/watch?v=1eTlYi3d9Ho.

7. Maywa Montenegro De Wit, "What Grows from a Pandemic? Toward an Abolitionist Agroecology," *Journal of Peasant Studies* 48, no. 1 (January 2, 2021): 100, https://doi.org/10.1080/03066150.2020.1854741.

8. Montenegro De Wit, 111–12.

9. Million Belay and Bridget Mugambe, "Bill Gates Should Stop Telling Africans What Kind of Agriculture Africans Need," *Scientific American*, July 6, 2021, https://www.scientificamerican.com/article/bill-gates-should-stop-telling-africans-what-kind-of-agriculture-africans-need1/.

10. Author interview, Monsanto Company, St. Louis, MO, May 27, 2015.

11. Author interview, Monsanto Company, St. Louis, MO, May 26, 2015.

12. Susan Reidy, "USDA, AGRA Partner to Help African Farmers," World-grain.com, March 31, 2022, https://www.world-grain.com/articles/16711-usda-agra-partner-to-help-african-farmers.

13. "Amcham Agricultural Symposium 2021," American Chamber of Commerce, Nairobi, Kenya. https://amchamke.glueup.com/event/amcham-agriculture-symposium-2021-43228/.

14. Office of the United States Trade Representative, "United States and Kenya Announce the Launch of the US-Kenya Strategic Trade and Investment Partnership," Office of the United States Trade Representative, July 14, 2022, https://ustr.gov/about-us/policy-offices/press-office/press-releases/2022/july/united-states-and-kenya-announce-launch-us-kenya-strategic-trade-and-investment-partnership.

15. Jesse Goldstein, *Planetary Improvement: Cleantech Entrepreneurship and the Contradictions of Green Capitalism*. (Cambridge, MA: MIT Press, 2018).

16. Julie Deering, "Who Owns CRISPR?," Seed World Canada, November 16, 2018, https://germination.ca/who-owns-crispr/; Maywa Montenegro de Wit, "Can Agroecology and CRISPR Mix? The Politics of Complementarity and Moving toward Technology Sovereignty," *Agriculture and Human Values* 39, no. 2 (June 1, 2022): 733–55, https://doi.org/10.1007/s10460-021-10284-0.

17. Tamar Haspell, "The Last Thing Africa Needs to Be Debating Is GMOs," *Washington Post*, May 22, 2015, https://www.washingtonpost.com/lifestyle/food/the-last-thing-africa-needs-to-be-debating-is-gmos/2015/05/22/81b76574-fe62-11e4-833c-a2de05b6b2a4_story.html?utm_term=.845e5f82a430.

18. The White House, "U.S. Strategy toward sub-Saharan Africa," https://www.whitehouse.gov/wp-content/uploads/2022/08/U.S.-Strategy-Toward-Sub-Saharan-Africa-FINAL.pdf.

19. Nosmot Gbadamosi, "New U.S. Africa Strategy Heavy on Competition with Russia, China," *Foreign Policy*, August 10, 2022, https://foreignpolicy.com/2022/08/10/us-africa-strategy-blinken-biden-thomas-greenfield-russia-china/.

20. Alex Thurston, "Biden's New Africa Strategy Is Shortsighted and Stale," *Responsible Statecraft*, August 12, 2022, https://responsiblestatecraft.org/2022/08/12/bidens-new-africa-strategy-is-shortsighted-and-stale/.

21. "How the Gates Foundation Is Driving the Food System, in the Wrong Direction," *Grain*, June 17, 2021, https://grain.org/en/article/6690-how-the-gates-foundation-is-driving-the-food-system-in-the-wrong-direction.

22. Bill Gates, *How to Avoid a Climate Disaster: The Solutions We Have and the Breakthroughs We Need* (New York: Knopf, 2021), 110.

23. Nick Cullather, *The Hungry World: America's Cold War Battle against Poverty in Asia* (Cambridge, MA: Harvard University Press, 2013).

24. Charles C. Mann, *The Wizard and the Prophet: Two Remarkable Scientists and Their Dueling Visions to Shape Tomorrow's World* (New York: Vintage, 2018).

25. Bill Gates, "Who Will Suffer Most from Climate Change? (Hint: Not You)," *GatesNotes* (blog), accessed January 12, 2021, https://www.gatesnotes.com/Energy/Who-Will-Suffer-Most-From-Climate-Change.

Bibliography

ARCHIVAL SOURCES

Rockefeller Archive Center, Tarrytown, New York
  Rockefeller Foundation Records
University of Minnesota Archives, Minneapolis, MN
  Norman Borlaug Papers
  E. C. Stakman Papers

U.S. GOVERNMENT PUBLICATIONS

Global Food Security Act of 2016, H.R. 1567, 114th Congress.
National Intelligence Council, *Global Trends: A Paradox of Progress*, 2017.
———. "Implications for US National Security of Anticipated Climate Change." September 2016.
———. "Intelligence Community Assessment: Global Food Security." September 2015.
U.S. Agency for International Development. "U.S. Government Global Food Security Strategy, FY 2017–2021." https://www.usaid.gov/sites/default/files/Global-Food-Security-Strategy_2017-2021.pdf.
———. "Why Resilience?" Video. September 30, 2021. https://www.usaid.gov/resilience.

BIBLIOGRAPHY

PRIMARY AND SECONDARY SOURCES

AATF. "Press Release: AATF Receives Grant to Make New Drought-Tolerant and Insect-Resistant Maize Hybrids Available to Farmers in Africa." June 18, 2018. https://www.aatf-africa.org/wp-content/uploads/2018/11/Press-release-Gates-Foundation-Grants-AATF-24m.pdf.

———. "TELA Maize Technology: FAQs." February 2021, https://www.aatf-africa.org/wp-content/uploads/2021/02/TELA-Project-FAQ.pdf.

Adas, Michael. *Machines as the Measure of Men: Science, Technology, and Ideologies of Western Dominance.* Ithaca, NY: Cornell University Press, 2014.

African Agricultural Technology Foundation. "Concept Note: Water Efficient Maize for Africa." Accessed June 19, 2015. http://www.aatf-africa.org.

Al-Bulushi, Samar. "Geographies of War-Making in East Africa," *Africa is a Country*, April 15, 2021. https://africasacountry.com/2021/04/geographies-of-war-making-in-east-africa.

Alliance for a Green Revolution in Africa. *Seeding an African Green Revolution: The PASS Journey.* Nairobi, Kenya, 2017.

"Amcham Agricultural Symposium 2021." American Chamber of Commerce, Nairobi, Kenya. https://amchamke.glueup.com/event/amcham-agriculture-symposium-2021-43228/.

Amseed. "U.S. Seed Industry Has a Role to Play in Sub-Saharan Africa." June 21, 2011. http://www.amseed.org.

Appel, Hannah. *The Licit Life of Capitalism: US Oil in Equatorial Guinea.* Durham NC: Duke University Press, 2019.

Arvin, Maile. *Possessing Polynesians: The Science of Settler Colonial Whiteness in Hawai'i and Oceania.* Durham, NC: Duke University Press, 2019.

Aschoff, Nicole Marie. *The New Prophets of Capital.* London: Verso, 2015.

Baranski, Marci. *The Globalization of Wheat: A Critical History of the Green Revolution.* Pittsburgh, PA: University of Pittsburgh Press, 2022.

Belay, Million, and Bridget Mugambe. "Bill Gates Should Stop Telling Africans What Kind of Agriculture Africans Need." *Scientific American*, July 6, 2021. https://www.scientificamerican.com/article/bill-gates-should-stop-telling-africans-what-kind-of-agriculture-africans-need1/.

Bender, Ruth. "How Bayer-Monsanto Became One of the Worst Corporate Deals—in 12 Charts." *Wall Street Journal (Online)*, August 28, 2019, sec. Business. https://www.proquest.com/docview/2281135814/citation/3988E25F879E49D8PQ/3.

Ben-zvi, Yael. "Where Did Red Go? Lewis Henry Morgan's Evolutionary Inheritance and U.S. Racial Imagination." *CR: The New Centennial Review* 7, no. 2 (2007): 201–29. https://doi.org/10.1353/ncr.2007.0037.

Bhandar, Brenna. *Colonial Lives of Property: Law, Land, and Racial Regimes of Ownership.* Durham, NC: Duke University Press, 2018.

Bird, Jane. "'Smart' Insurance Helps Poor Farmers to Cut Risk." December 5, 2018. https://www.ft.com/content/3a8c7746-d886-11e8-aa22-3653848 7e3do.

Bjerga, Alan. "Sowing the Seeds of a Farm Boom in Africa." *Bloomberg*, March 31, 2016. https://www.bloomberg.com/news/articles/2016-03-31 /sowing-the-seeds-of-a-farm-boom-in-africa.

Black, Megan. *The Global Interior: Mineral Frontiers and American Power*. Cambridge, MA: Harvard University Press, 2018.

Bolton, John R. "Remarks by National Security Advisor Ambassador John R. Bolton on the Trump Administration's New Africa Strategy." The White House. Accessed September 14, 2020. https://www.whitehouse.gov/briefings -statements/remarks-national-security-advisor-ambassador-john-r-bolton -trump-administrations-new-africa-strategy/.

Borlaug, Norman E. "Ending World Hunger: The Promise of Biotechnology and the Threat of Antiscience Zealotry." *Plant Physiology* 124, no. 2 (2000): 487–90.

———. "Feeding a Hungry World." *Science* 318, no. 5849 (October 19, 2007): 359–359. https://doi.org/10.1126/science.1151062.

———. "A Green Revolution for Africa." *Wall Street Journal*, October 26, 2007, sec. Opinion. https://www.wsj.com/articles/SB119336762148772617.

———. Mankind and Civilization at Another Crossroad." Presented at the FAO, Rome, Italy, November 8, 1971. https://www.fao.org/3/c3017e/c3017e.pdf.

———. "Mobilizing Science and Technology for a Green Revolution in Achieving Greater Impact from Research Investments in Africa." Edited by Steven Breth. Mexico City: Sasakawa Africa Association, 1996.

Boyd, William. "Wonderful Potencies? Deep Structure and the Problem of Monopoly in Agricultural Biotechnology." In *Engineering Trouble: Biotechnology and Its Discontents*, edited by Rachel A. Schurman and Dennis Doyle Takahashi Kelso, 24–62. Berkeley: University of California Press, 2003.

Bronson, Kelly. "The Dangers of Big Data Extend to Farming." *The Conversation*, June 27, 2022. http://theconversation.com/the-dangers-of-big-data-extend-to -farming-184531.

Bruyneel, Kevin. *Settler Memory: The Disavowal of Indigeneity and the Politics of Race in the United States*. Chapel Hill: University of North Carolina Press, 2021.

Burger, Ludwig, and Patricia Weiss. "Bayer's Agriculture Unit, Consumer Health Drive Outlook Hike." *Reuters*, August 4, 2022. https://www.reuters .com/business/healthcare-pharmaceuticals/bayers-agriculture-unit -consumer-health-drives-improved-outlook-2022-2022-08-04/.

Byerlee, Derek. "The Globalization of Hybrid Maize, 1921–70." *Journal of Global History* 15, no. 1 (March 2020): 101–22.

Byerlee, Derek, and Greg Edmeades. *Fifty Years of Maize Research in the CGIAR*. Mexico City: CIMMYT, 2021. https://doi.org/10.1017/S1740022819000354.

Canfield, Matthew. "The Ideology of Innovation: Philanthropy and Racial Capitalism in Global Food Governance." *Journal of Peasant Studies*, September 19, 2022, 1–25. https://doi.org/10.1080/03066150.2022.2099739.

Chakravartty, Paula, and Denise Ferreira Da Silva. "Accumulation, Dispossession, and Debt: The Racial Logic of Global Capitalism—An Introduction." *American Quarterly* 64, no. 3 (2012): 361–85. https://doi.org/10.1353/aq.2012.0033.

Chambers, Samuel A. *There's No Such Thing as the Economy: Essays on Capitalist Value*. Santa Barbara, CA: Punctum Books, 2018.

Chang, David A. *The Color of the Land: Race, Nation, and the Politics of Landownership in Oklahoma, 1832–1929*. Chapel Hill, N.C: The University of North Carolina Press, 2010.

CIMMYT. "CRISPR-CAS Technology by Neal Guterson [sic]." 2016. https://www.youtube.com/watch?v=goOmHFQyoXI&list=PLjXdzeDP_y5FEhl8XKm7dosQbArng7mn3&index=21.

———. "DuPont Pioneer and CIMMYT Form CRISPR-Cas Public/Private Partnership." CIMMYT, September 28, 2016. http://www.cimmyt.org/press_release/dupont-pioneer-and-cimmyt-form-crispr-cas-publicprivate-partnership/Sept. 28.

———. "New Screening Cycle Begins for Maize Lethal Necrosis Disease in Kenya." October 3, 2017. https://www.cimmyt.org/news/new-screening-cycle-for-deadly-mln-virus-set-to-begin-in-kenya/.

Connelly, Matthew. *Fatal Misconception: The Struggle to Control World Population*. Cambridge. MA: Harvard University Press, 2010.

Cooper, Melinda. "Turbulent Worlds: Financial Markets and Environmental Crisis." *Theory, Culture & Society* 27, nos. 2–3 (March 1, 2010): 167–90. https://doi.org/10.1177/0263276409358727.

Cotter, Joseph. *Troubled Harvest: Agronomy and Revolution in Mexico, 1880–2002*. Westport, CT: Praeger, 2003.

CSIS. "The Power of Global Food Security: Examining Economic and National Security Implications." 2016. https://www.csis.org/events/power-global-food-security-examining-economic-and-national-security-implications.

Cullather, Nick. *The Hungry World: America's Cold War Battle against Poverty in Asia*. Cambridge, MA: Harvard University Press, 2013.

———. "Miracles of Modernization: The Green Revolution and the Apotheosis of Technology." *Diplomatic History* 28, no. 2 (2004): 227–54.

Curry, Helen Anne. "Taxonomy, Race Science, and Mexican Maize." *Isis* 112, no. 1 (March 1, 2021): 1–21. https://doi.org/10.1086/713819.

Deering, Julie. "Who Owns CRISPR?" Seed World Canada, November 16, 2018. https://germination.ca/who-owns-crispr/.

Deloria, Philip Joseph, and Alexander I. Olson. *American Studies: A User's Guide*. Oakland: University of California Press, 2017.

Democracy Now! "The Global Climate Wall: Wealthy Nations Prioritize Militarizing Borders over Climate Action." November 10, 2021. https://www.democracynow.org/2021/11/10/global_climate_wall.
Dhamoon, Rita. "A Feminist Approach to Decolonizing Anti-Racism: Rethinking Transnationalism, Intersectionality, and Settler Colonialism." *Feral Feminisms* 4 (summer 2015). https://feralfeminisms.com/rita-dhamoon/.
Doering, Don S. "Public-Private Partnership to Develop and Deliver Drought Tolerant Crops to Food-Insecure Farmers." Summary and Interpretation of the May 3-4 Strategy and Planning Meeting. Winrock International, May 31, 2005.
Domonoske, Camila. "Monsanto No More: Agri-Chemical Giant's Name Dropped in Bayer Acquisition." *NPR*, June 4, 2018, sec. America. https://www.npr.org/sections/thetwo-way/2018/06/04/616772911/monsanto-no-more-agri-chemical-giants-name-dropped-in-bayer-acquisition.
Dunbar-Ortiz, Roxanne. *An Indigenous Peoples' History of the United States.* Boston: Beacon Press, 2014.
Dyer, Richard. *White*. London: Routledge, 1997.
Economist. "In Africa, Agricultural Insurance Often Falls on Stony Ground." December 15, 2018. https://www.economist.com/finance-and-economics/2018/12/15/in-africa-agricultural-insurance-often-falls-on-stony-ground.
Eddens, Aaron. "White Science and Indigenous Maize: The Racial Logics of the Green Revolution." *Journal of Peasant Studies* 46, no. 3 (April 16, 2019): 653–73. https://doi.org/10.1080/03066150.2017.1395857.
Ehrlich, Paul. *The Population Bomb*. New York: Ballentine, 1968.
Elmore, Bartow J. *Seed Money: Monsanto's Past and Our Food Future.* New York: W. W. Norton & Company, 2021.
Essex, Jamey. *Development, Security, and Aid: Geopolitics and Geoeconomics at the U.S. Agency for International Development.* Athens: University of Georgia Press, 2013.
Federici, Silvia. *Re-enchanting the World: Feminism and the Politics of the Commons.* Oakland, CA: PM Press, 2019.
Feldman, Keith P. "The Globality of Whiteness in Post-Racial Visual Culture." *Cultural Studies* 30, no. 2 (March 3, 2016): 289–311. https://doi.org/10.1080/09502386.2015.1020957.
Ferguson, Roderick A. *The Reorder of Things: The University and Its Pedagogies of Minority Difference.* Minneapolis: University of Minnesota Press, 2012.
Financial Times. "De-Risking Agricultural Investment in Africa." August 14, 2018. https://www.ft.com/content/4ee682ec-9fd6-11e8-85da-eeb7a9ce36e4.
Fitting, Elizabeth M. *The Struggle for Maize: Campesinos, Workers, and Transgenic Corn in the Mexican Countryside.* Durham, NC: Duke University Press, 2011.

Fitzgerald, Deborah. "Exporting American Agriculture: The Rockefeller Foundation in Mexico, 1943–53." *Social Studies of Science* 16, no. 3 (1986): 457–83.

Flavelle, Christopher, Julian E. Barnes, Eileen Sullivan, and Jennifer Steinhauer. "Climate Change Poses a Widening Threat to National Security." *New York Times*, October 21, 2021, sec. Climate. https://www.nytimes.com/2021/10/21/climate/climate-change-national-security.html.

Foucault, Michel. "Nietzsche, Genealogy, History." In *The Foucault Reader*, edited by Paul Rabinow, 76–100. New York: Pantheon, 1984.

Fry, Suzanne. "National Security Is Food Security: Strategic Leadership and a Moral Imperative: Panel Remarks." Presented at the Global Food Security Symposium, Washington, D.C., March 30, 2017.

Gabay, Clive. *Imagining Africa: Whiteness and the Western Gaze*. Cambridge: Cambridge University Press, 2018.

Gabor, Daniela, and Sally Brooks. "The Digital Revolution in Financial Inclusion: International Development in the Fintech Era." *New Political Economy* 22, no. 4 (July 4, 2017): 423–36. https://doi.org/10.1080/13563467.2017.1259298.

Gates, Bill. "Helping Poor Farmers Grow Their Crops." *GatesNotes* (blog), January 24, 2012. https://www.gatesnotes.com/The-Man-Who-Fed-the-World.

———. *How to Avoid a Climate Disaster: The Solutions We Have and the Breakthroughs We Need*. New York: Knopf, 2021.

———. "Support for the World's Poorest Farmers." World Food Prize, October 15, 2009. https://www.worldfoodprize.org/documents/filelibrary/images/borlaug_dialogue/2009_speakers/transcripts/2009BorlaugDialogueGatesbrief_65B2AF6BB5B25.pdf.

———. "Who Will Suffer Most from Climate Change? (Hint: Not You)." *GatesNotes* (blog). Accessed January 12, 2021. https://www.gatesnotes.com/Energy/Who-Will-Suffer-Most-From-Climate-Change.

Gbadamosi, Nosmot. "New U.S. Africa Strategy Heavy on Competition with Russia, China." *Foreign Policy*, August 10, 2022. https://foreignpolicy.com/2022/08/10/us-africa-strategy-blinken-biden-thomas-greenfield-russia-china/.

Gidwani, Vinay K. *Capital, Interrupted: Agrarian Development and the Politics of Work in India*. Minneapolis: University of Minnesota Press, 2008.

Gillis, Justin. "Can the Yield Gap Be Closed—Sustainably?" *New York Times* Green Blog, June 7, 2011. https://www.proquest.com/docview/2216870574/abstract/1882F1BC64ED4686PQ/1.

———. "Norman Borlaug, Father of a Crop Revolution, Dies at 95." *New York Times*, September 13, 2009, sec. Energy & Environment. https://www.nytimes.com/2009/09/14/business/energy-environment/14borlaug.html.

Glover, Dominic. "The Corporate Shaping of GM Crops as a Technology for the Poor." *Journal of Peasant Studies* 37, no. 1 (January 2010): 67–90. https://doi.org/10.1080/03066150903498754.

Glover, Kaiama L. "'Flesh Like One's Own': Benign Denials of Legitimate Complaint." *Public Culture* 29, no. 2 (May 1, 2017): 235–60. https://doi.org/10.1215/08992363-3749045.

Goldberg, David Theo. *Are We All Postracial Yet?* Malden, MA: Polity Press, 2015.

Goldstein, Jesse. *Planetary Improvement: Cleantech Entrepreneurship and the Contradictions of Green Capitalism.* Cambridge, MA: MIT Press, 2018.

Graddy, T. Garrett. "Situating In Situ: A Critical Geography of Agricultural Biodiversity Conservation in the Peruvian Andes and Beyond." *Antipode* 46, no. 2 (March 2014): 426–54. https://doi.org/10.1111/anti.12045.

Graddy-Lovelace, Garrett. "The Coloniality of US Agricultural Policy: Articulating Agrarian (in)Justice." *Journal of Peasant Studies* 44, no. 1 (January 2, 2017): 78–99. https://doi.org/10.1080/03066150.2016.1192133.

Grain. "How the Gates Foundation Is Driving the Food System, in the Wrong Direction." June 17, 2021. https://grain.org/en/article/6690-how-the-gates-foundation-is-driving-the-food-system-in-the-wrong-direction.

Greatrex, Helen, James Hansen, Samanth Garven, Rachel Diro, Sari Blakeley, and Margot Le Guen. "Scaling Up Index Insurance for Smallholder Farmers: Recent Evidence and Insights." CCAFS Working Paper, 2015. www.ccafs.cgiar.org.

Grewal, Inderpal. *Saving the Security State: Exceptional Citizens in Twenty-First-Century America.* Durham, NC: Duke University Press, 2017.

Hall, Stuart. "The West and the Rest: Discourse and Power." In *Modernity: An Introduction to Modern Societies,* edited by Stuart Hall, David Held, Don Hubert, and Kenneth Thompson, 184–227. Malden, MA: Blackwell, 1996.

Haraway, Donna. *Modest_Witness@Second_Millennium.FemaleMan_Meets_OncoMouse: Feminism and Technoscience.* New York: Routledge, 1997.

Haspell, Tamar. "The Last Thing Africa Needs to Be Debating Is GMOs." *Washington Post,* May 22, 2015. https://www.washingtonpost.com/lifestyle/food/the-last-thing-africa-needs-to-be-debating-is-gmos/2015/05/22/81b76574-fe62-11e4-833c-a2de05b6b2a4_story.html?utm_term=.845e5f82a430.

Herdt, Robert W. "People, Institutions, and Technology: A Personal View of the Role of Foundations in International Agricultural Research and Development 1960–2010." *Food Policy* 37, no. 2 (April 2012): 179–90. https://doi.org/10.1016/j.foodpol.2012.01.003.

Hesser, Leon F. *The Man Who Fed the World: Nobel Peace Prize Laureate Norman Borlaug and His Battle to End World Hunger: An Authorized Biography.* Dallas, TX: Durban House, 2006.

Hewitt de Alcantara, Cynthia. *Modernization of Mexican Agriculture.* Geneva: UNRISD, 1976.

Hildyard, Nicholas. "'Scarcity' as Political Strategy: Reflections on Three Hanging Children." In *The Limits to Scarcity*, edited by Lyla Mehta. London: Earthscan, 2010.

Holleman, Hannah. *Dust Bowls of Empire: Imperialism, Environmental Politics, and the Injustice of "Green" Capitalism*. New Haven, CT: Yale University Press, 2018.

Howard, Philip. "Recent Changes in the Global Seed Industry and Digital Agriculture Industries." *Philip H. Howard.net* (blog), January 4, 2023. https://philhoward.net/2023/01/04/seed-digital/.

Immerwahr, Daniel. *How to Hide an Empire: A History of the Greater United States*. Reprint edition. New York: Picador, 2020.

Isakson, S. Ryan. "Derivatives for Development? Small-Farmer Vulnerability and the Financialization of Climate Risk Management." *Journal of Agrarian Change* 15, no. 4 (2015): 569–80. https://doi.org/10.1111/joac.12124.

Jasanoff, Sheila, ed. *States of Knowledge: The Co-production of Science and Social Order*. London: Routledge, 2010.

Jennings, Bruce H. *Foundations of International Agricultural Research: Science and Politics in Mexican Agriculture*. Boulder, CO: Westview Press, 1988.

Jodi A. Byrd. *The Transit of Empire: Indigenous Critiques of Colonialism*. Minneapolis: University of Minnesota Press, 2011.

Johnson, Colin R. *Just Queer Folks: Gender and Sexuality in Rural America*. Philadelphia: Temple University Press, 2013.

Johnson, Leigh. "Index Insurance and the Articulation of Risk-Bearing Subjects." *Environment & Planning A* 45, no. 11 (2013): 2663–81. https://doi.org/10.1068/a45695.

Juma, Calestous. "How to Improve Africa's Seed Industry." World Economic Forum, September 11, 2015. https://www.weforum.org/agenda/2015/09/how-to-improve-africas-seed-industry/.

Jung, Moon-Ho. *Menace to Empire: Anticolonial Solidarities and the Transpacific Origins of the US Security State*. Oakland: University of California Press, 2023.

Kaplan, Amy. "'Left Alone with America': The Absence of Empire in the Study of American Culture." In *Cultures of United States Imperialism*, edited by Amy Kaplan and Donald E Pease. Durham, NC: Duke University Press, 1993.

Kaplan, Amy, and Donald E. Pease, eds. *Cultures of United States Imperialism*. Durham, NC: Duke University Press, 1993.

Kerry, John. "Remarks at a Working Session on Resilience and Food Security in a Changing Climate." U.S. State Department, August 4, 2014. https://20092017.state.gov/secretary/remarks/2014/08/230219.htm.

Kimmelman, Barbara A. "The American Breeders' Association: Genetics and Eugenics in an Agricultural Context, 1903–13." *Social Studies of Science* 13, no. 2 (May 1, 1983): 163–204. https://doi.org/10.1177/03063128301300200.

Kinkela, David. *DDT and the American Century: Global Health, Environmental Politics, and the Pesticide That Changed the World*. Chapel Hill: University of North Carolina Press, 2013.
Kinver, Mark. "Lack of Seeds Hampers Africa's Ability to Boost Yields." BBC News, March 10, 2016, sec. Science & Environment. https://www.bbc.com/news/science-environment-35774445.
Kish, Zenia, and Justin Leroy. "Bonded Life: Technologies of Racial Finance from Slave Insurance to Philanthrocapital." *Cultural Studies* 29, nos. 5–6 (September 3, 2015): 630–51. https://doi.org/10.1080/09502386.2015.1017137.
Klein, Naomi. *This Changes Everything: Capitalism vs. the Climate*. New York: Simon & Schuster, 2014.
Kloppenburg, Jack R. Jr. *First the Seed: The Political Ecology of Plant Biotechnology, 1492–2000*. 2nd edition. Madison: University of Wisconsin Press, 2004.
Knobloch, Frieda. *The Culture of Wilderness*. Chapel Hill: University of North Carolina Press, 1996.
Koigi, Bob. "Corteva Agriscience Launches Operations in East Africa." Africa Business Communities, March 9, 2019. https://africabusinesscommunities.com/news/corteva-agriscience-launches-operations-in-east-africa/.
Landler, Mark. "Curing the Ills of America's Top Foreign Aid Agency." *New York Times*, October 23, 2010, sec. World. https://www.nytimes.com/2010/10/23/world/23shah.html.
Lapavitsas, Costas. *Profiting without Producing: How Finance Exploits Us All*. London: Verso, 2013.
Lee, Christopher, and Melani McAlister. "Introduction: Generations of Empire in American Studies." *American Quarterly* 74, no. 3 (2022): 477–97.
Lee, Robert, and Tristan Ahtone. "Land-Grab Universities." *High Country News*, March 30, 2020. https://www.hcn.org/issues/52.4/indigenous-affairs-education-landgrab-universities.
Levy, Jonathan. *Freaks of Fortune: The Emerging World of Capitalism and Risk in America*. Cambridge, MA: Harvard University Press, 2012.
Linebaugh, Peter, and Marcus Rediker. *The Many-Headed Hydra: Sailors, Slaves, Commoners, and the Hidden History of the Revolutionary Atlantic*. Reprint edition. Boston: Beacon Press, 2013.
"LIVE: CNBC's Steve Sedgwick Joins Davos Panel on Revolutionizing Food Security—1/18/23." Video, 2023. https://www.youtube.com/watch?v=1eTlYi3d9H0.
Locke, John. *The Second Treatise of Government and A Letter Concerning Toleration*. Mineola, NY: Dover, 2002.
Lowe, Lisa. *The Intimacies of Four Continents*. Durham, NC: Duke University Press, 2015.
Luna, Jessie K. "The Chain of Exploitation: Intersectional Inequalities, Capital Accumulation, and Resistance in Burkina Faso's Cotton Sector." *Journal of*

Peasant Studies 46, no. 7 (November 10, 2019): 1413–34. https://doi.org/10.1080/03066150.2018.1499623.

Lutz, Catherine, and Jane Lou Collins. *Reading National Geographic*. Chicago: University of Chicago Press, 1993.

Mahuku, George, Benham E. Lockhart, Bramwel Wanjala, Mark W. Jones, Janet Njeri Kimunye, Lucy R. Stewart, Bryan J. Cassone, et al. "Maize Lethal Necrosis (MLN), an Emerging Threat to Maize-Based Food Security in Sub-Saharan Africa." *Phytopathology* 105, no. 7 (March 30, 2015): 956–65. https://doi.org/10.1094/PHYTO-12-14-0367-FI.

Mann, Charles C. *The Wizard and the Prophet: Two Remarkable Scientists and Their Dueling Visions to Shape Tomorrow's World*. New York: Vintage, 2018.

Martin, Randy. *An Empire of Indifference: American War and the Financial Logic of Risk Management*. Durham, NC: Duke University Press, 2007.

Marx, Karl. *Capital: A Critique of Political Economy, Volume 1*. Translated by Ben Fowkes. Reprint edition. London: Penguin Classics, 1992.

Mayersohn, Norman. "He Grew Up on a Farm; Now, He Helps Protect Them," *New York Times*, October 3, 2019. https://www.nytimes.com/2019/10/03/business/microinsurance-africa-thomas-njeru.html.

McCann, James. *Maize and Grace: Africa's Encounter with a New World Crop, 1500–2000*. Cambridge, MA: Harvard University Press, 2005.

McGoey, Linsey. *No Such Thing as a Free Gift: The Gates Foundation and the Price of Philanthropy*. London: Verso, 2016.

McKibben, Bill. "How Does Bill Gates Plan to Solve the Climate Crisis?" *New York Times*, February 15, 2021. https://www.proquest.com/docview/2489142477/citation/5D6B62DAB77D4324PQ/1.

McKittrick, Katherine. "Plantation Futures." *Small Axe: A Caribbean Journal of Criticism* 17, no. 3 (January 1, 2013): 1–15. https://doi.org/10.1215/07990537-2378892.

Mehta, Lyla, ed. *The Limits to Scarcity: Contesting the Politics of Allocation*. London: Earthscan, 2010.

Melamed, Jodi. "Racial Capitalism." *Critical Ethnic Studies* 1, no. 1 (2015): 76. https://doi.org/10.5749/jcritethnstud.1.1.0076.

Mills, Charles W. "Global White Ignorance." In *Routledge International Handbook of Ignorance Studies*. London: Routledge, 2015.

———. "White Ignorance." In *Race and Epistemologies of Ignorance*, edited by Shannon Sullivan and Nancy Tuana, 13–38. Albany: State University of New York Press, 2007.

Monsanto/AATF. "Combining Breeding and Biotechnology to Develop Drought Tolerant Maize for Africa: A Proposal to the Bill and Melinda Gates Foundation." May 25, 2007, St. Louis.

Montenegro, Maywa. "Opinion: CRISPR Is Coming to Agriculture—with Big Implications for Food, Farmers, Consumers and Nature." *Ensia*, January 28,

2016. https://ensia.com/voices/crispr-is-coming-to-agriculture-with-big-implications-for-food-farmers-consumers-and-nature/.

Montenegro de Wit, Maywa. "Can Agroecology and CRISPR Mix? The Politics of Complementarity and Moving toward Technology Sovereignty." *Agriculture and Human Values* 39, no. 2 (June 1, 2022): 733–55. https://doi.org/10.1007/s10460-021-10284-0.

———. "Stealing into the Wild: Conservation Science, Plant Breeding and the Makings of New Seed Enclosures." *Journal of Peasant Studies* 44, no. 1 (January 2, 2017): 169–212. https://doi.org/10.1080/03066150.2016.1168405.

———. "What Grows from a Pandemic? Toward an Abolitionist Agroecology." *Journal of Peasant Studies* 48, no. 1 (January 2, 2021): 99–136. https://doi.org/10.1080/03066150.2020.1854741.

Moreton-Robinson, Aileen. *The White Possessive: Property, Power, and Indigenous Sovereignty*. Minneapolis: University of Minnesota Press, 2015.

Moseley, William G., and Melanie Ouedraogo. "When Agronomy Flirts with Markets, Gender, and Nutrition: A Political Ecology of the New Green Revolution for Africa and Women's Food Security in Burkina Faso." *African Studies Review* 65, no. 1 (March 2022): 41–65. https://doi.org/10.1017/asr.2021.74.

MPR News. "Norman Borlaug Statue Unveiled at US Capitol." MPR News, March 25, 2014. https://www.mprnews.org/story/2014/03/25/news/borlaug-statue.

Mukherjee, Roopali. "Antiracism Limited: A Pre-History of Post-Race." *Cultural Studies* 30, no. 1 (January 2, 2016): 47–77. https://doi.org/10.1080/09502386.2014.935455.

Mukherjee, Roopali, Sarah Banet-Weiser, and Herman Gray, eds. *Racism Postrace*. Durham: Duke University Press, 2019.

Nader, Laura. "Up the Anthropologist: Perspectives Gained from Studying Up." In *Reinventing Anthropology*, edited by Dell Hymes. New York: Random House, 1969.

Nanteza, Winnie. "WEMA Achieves Major Milestone in African Agriculture." Alliance for Science, May 29, 2018. https://allianceforscience.org/blog/2018/05/wema-achieves-major-milestone-african-agriculture/.

Newman, Amie. "Gates Foundation Visitor Center: An Interview with Therese Littleton." Impatient Optimists, February 4, 2012. https://www.impatientoptimists.org/Posts/2012/02/Use-Your-Voice-for-Good-An-Interview-with-Therese-Littleton.

O'Brien, Jean M. *Firsting and Lasting: Writing Indians out of Existence in New England*. Minneapolis: University of Minnesota Press, 2010.

Office of the United States Trade Representative. "United States and Kenya Announce the Launch of the U.S.-Kenya Strategic Trade and Investment Partnership." July 14, 2022. https://ustr.gov/about-us/policy-offices/press

-office/press-releases/2022/july/united-states-and-kenya-announce-launch-us-kenya-strategic-trade-and-investment-partnership.

Olsson, Tore C. *Agrarian Crossings: Reformers and the Remaking of the US and Mexican Countryside*. Princeton, NJ: Princeton University Press, 2017.

Onditi, Geoffrey. "Saturday Morning Interview (Julie Borlaug) KBC." YouTube, September 19, 2016. https://www.youtube.com/watch?v=VoFZWofCfEs.

Paarlberg, Robert L. *Starved for Science: How Biotechnology Is Being Kept Out of Africa*. Cambridge, MA: Harvard University Press, 2009.

Patel, Raj. "The Long Green Revolution." *Journal of Peasant Studies* 40, no. 1 (January 2013): 1–63. https://doi.org/10.1080/03066150.2012.719224.

Patel, Raj, and Jason W Moore. *A History of the World in Seven Cheap Things: A Guide to Capitalism, Nature, and the Future of the Planet*. Oakland: University of California Press, 2018.

Perez, Enrique. "Collective Action, Democracy and Mexico's Defense of Its Corn and Food Sovereignty." Institute for Agriculture and Trade Policy, December 20, 2022. https://www.iatp.org/collective-action-democracy-mexicos-defense-corn-food-sovereignty.

Perkins, John H. *Geopolitics and the Green Revolution: Wheat, Genes, and the Cold War*. New York: Oxford University Press, 1997.

Philpott, Tom. "Bill Gates Reveals Support for GMO Ag." *Grist*, October 22, 2009. https://grist.org/article/2009-10-21-bill-gates-reveals-support-for-gmo-ag/.

Pioneer. "History of Pioneer." https://www.pioneer.com/us/about-us/our-history.html.

Pray, Carl E., and Ruben G. Echeverria. "Transferring Hybrid Maize Technology: The Role of the Private Sector." *Food Policy* 13, no. 4 (November 1, 1988): 366–74. https://doi.org/10.1016/0306-9192(88)90084-X.

Pulido, Laura. "Racism and the Anthropocene." In *Future Remains: A Cabinet of Curiosities for the Anthropocene*, edited by Gregg Mitman, Marco Armiero, and Robert S. Emmett, 116–28. Chicago: University of Chicago Press, 2018.

Quinn, Kenneth. "Extended Biography: Norman E. Borlaug." World Food Prize, 2009. https://www.worldfoodprize.org/en/dr_norman_e_borlaug/extended_biography/.

———. "Quinn: A Tribute to Norman Borlaug on the Fourth Anniversary of His Death." World Food Prize. https://www.worldfoodprize.org/index.cfm/87428/40197/quinn_a_tribute_to_norman_borlaug_on_the_fourth_anniversary_of_his_death.

Reidy, Susan. "USDA, AGRA Partner to Help African Farmers," World-grain.com, March 31, 2022, https://www.world-grain.com/articles/16711-usda-agra-partner-to-help-african-farmers.

Rock, Joeva Sean. *We Are Not Starving: The Struggle for Food Sovereignty in Ghana*. Lansing: Michigan State University Press, 2022.

Rockefeller Foundation. "Africa's Turn: A New Green Revolution for the 21st Century." July 2006. https://assets.rockefellerfoundation.org/app

/uploads/20060701123216/dc8aefda-bc49-4246-9e92-9026bc0eed04-africas_turn.pdf.

Rockefeller Foundation. *Director's Annual Report: September 1958–August 1959: Mexican Agriculture Program.* New York: The Rockefeller Foundation, 1959.

Rodney, Walter. *How Europe Underdeveloped Africa.* Washington, DC: Howard University Press, 1981.

Rosenberg, Gabriel N. "No Scrubs: Livestock Breeding, Eugenics, and the State in the Early Twentieth-Century United States." *Journal of American History* 107, no. 2 (September 1, 2020): 362–87. https://doi.org/10.1093/jahist/jaaa179.

Rosenberg, Tina. "A Green Revolution, This Time for Africa," *New York Times*, Opinionater, April 9, 2014. https://archive.nytimes.com/opinionator.blogs.nytimes.com/2014/04/09/a-green-revolution-this-time-for-africa/.

Ross, Eric B. *The Malthus Factor: Population, Poverty, and Politics in Capitalist Development.* London: Zed Books, 1998.

Roy, Ananya. "Introduction: The Aporias of Poverty." In *Territories of Poverty*, edited by Ananya Roy and Emma Shaw Crane, 1–36. Athens: University of Georgia Press, 2015.

———. *Poverty Capital: Microfinance and the Making of Development.* New York: Routledge, 2010.

Saldaña-Portillo, María Josefina. *Indian Given: Racial Geographies across Mexico and the United States.* Durham, NC: Duke University Press, 2016.

Schnurr, Matthew A. *Africa's Gene Revolution: Genetically Modified Crops and the Future of African Agriculture.* Montreal: McGill-Queen's University Press, 2019.

———. "Biotechnology and Bio-Hegemony in Uganda: Unraveling the Social Relations Underpinning the Promotion of Genetically Modified Crops into New African Markets." *Journal of Peasant Studies* 40, no. 4 (July 2013): 639–58. https://doi.org/10.1080/03066150.2013.814106.

Schultz, Theodore W. *Transforming Traditional Agriculture.* New Haven, CT: Yale University Press, 1964.

Schurman, Rachel. "Building an Alliance for Biotechnology in Africa." *Journal of Agrarian Change* 17, no. 3 (July 2017): 441–58. https://doi.org/10.1111/joac.12167.

———. "Micro(Soft) Managing a 'Green Revolution' for Africa: The New Donor Culture and International Agricultural Development." *World Development* 112 (2018): 180–92. https://doi.org/10.1016/j.worlddev.2018.08.003.

Schurman, Rachel, and William A. Munro. *Fighting for the Future of Food: Activists versus Agribusiness in the Struggle over Biotechnology.* Minneapolis: University of Minnesota Press, 2010.

Schwab, Tim. *The Bill Gates Problem: Reckoning with the Myth of the Good Billionaire.* New York: Metropolitan Books, 2023.

Scott, James C. *Seeing Like a State: How Certain Schemes to Improve the Human Condition Have Failed.* New Haven, CT: Yale University Press, 2008.

Shepherd, Chris J. "Imperial Science: The Rockefeller Foundation and Agricultural Science in Peru, 1940–1960." *Science as Culture* 14, no. 2 (2005): 113–37.

Shulman, George M. *American Prophecy: Race and Redemption in American Political Culture*. Minneapolis: University of Minnesota Press, 2008.

Soto Laveaga, Gabriela. "Beyond Borlaug's Shadow: Octavio Paz, Indian Farmers, and the Challenge of Narrating the Green Revolution." *Agricultural History* 95, no. 4 (October 1, 2021): 576–608. https://doi.org/10.3098/ah.2021.095.4.576.

Squires, Catherine R. *The Post-Racial Mystique: Media and Race in the Twenty-First Century*. New York: New York University Press, 2014.

Sreenivas, Mytheli. *Reproductive Politics and the Making of Modern India*. Seattle: University of Washington Press, 2021.

Stakman, E. C, Richard Bradfield, and Paul C Mangelsdorf. *Campaigns against Hunger*. Cambridge, MA: Belknap Press, 1967.

Stone, Glenn Davis. *The Agricultural Dilemma: How Not to Feed the World*. New York: Routledge, 2022.

Sturken, Marita. *Tangled Memories: The Vietnam War, the Aids Epidemic, and the Politics of Remembering*. Berkeley: University of California Press, 1997.

Sumberg, James, Dennis Keeney, and Benedict Dempsey. "Public Agronomy: Norman Borlaug as 'Brand Hero' for the Green Revolution." *Journal of Development Studies* 48, no. 11 (November 2012): 1587–600. https://doi.org/10.1080/00220388.2012.713470.

TallBear, Kim. *Native American DNA: Tribal Belonging and the False Promise of Genetic Science*. Minneapolis: University of Minnesota Press, 2013.

Thurston, Alex. "Biden's New Africa Strategy Is Shortsighted and Stale." *Responsible Statecraft*, August 12, 2022. https://responsiblestatecraft.org/2022/08/12/bidens-new-africa-strategy-is-shortsighted-and-stale/.

Turse, Nick. "Exclusive: The U.S. Has More Military Operations in Africa than the Middle East." *Vice News*, December 12, 2018, https://www.vice.com/en_us/article/a3my38/exclusive-the-us-has-more-military-operations-in-africa-than-the-middle-east.

———. *Tomorrow's Battlefield: U.S. Proxy Wars and Secret Ops in Africa*. Chicago: Haymarket Books, 2015.

———. "The U.S. Military Moves Deeper into Africa." *TomDispatch*, April 27, 2017. http://www.tomdispatch.com/post/176272/tomgram%3A_nick_turse%2C_the_u.s._military_moves_deeper_into_africa/.

Turse, Nick, Sam Mednick, and Amanda Sperber. "Exclusive: Inside the Secret World of US Commandos in Africa." Pulitzer Center, August 11, 2020. https://pulitzercenter.org/reporting-exclusive-inside-secret-world-us-commandos-africa.

Verdery, Katherine. *The Vanishing Hectare: Property and Value in Postsocialist Transylvania*. Ithaca, NY: Cornell University Press, 2003.

Vernon, James. *Hunger: A Modern History*. Cambridge, MA: The Belknap Press of Harvard University Press, 2007.

Vine, David. *The United States of War*. Oakland: University of California Press, 2020.

Visser, Jaco. "Monsanto Targets Smallholder Farmers." *Farmer's Weekly*, January 8, 2015. https://www.farmersweekly.co.za/bottomline/monsanto-targets-smallholder-farmers/.

Walker, Jeremy, and Melinda Cooper. "Genealogies of Resilience: From Systems Ecology to the Political Economy of Crisis Adaptation." *Security Dialogue* 42, no. 2 (April 1, 2011): 143–60. https://doi.org/10.1177/0967010611399616.

Wall Street Journal. Editorial board. "Battering Norman Borlaug." *Wall Street Journal*, Eastern edition, April 25, 2020.

Waltz, Emily. "Beating the Heat." *Nature Biotechnology* 32, no. 7 (July 2014): 610–13. https://doi.org/10.1038/nbt.2948.

Wellhausen, E. J. "Exotic Germ Plasm for Improvement of Corn Belt Maize." In *Proceedings of the 20th Annual Hybrid Corn Industry-Research Conference*. Chicago, 1965.

———. *Races of Maize in Mexico: Their Origin, Characteristics and Distribution*. Cambridge, MA: Bussey Institution of Harvard University, 1952.

The White House. "Remarks by the President at Symposium on Global Agriculture and Food Security." May 18, 2012. https://obamawhitehouse.archives.gov/the-press-office/2012/05/18/remarks-president-symposium-global-agriculture-and-food-security.

———. "U.S. Strategy toward Sub-Saharan Africa." August 2022. https://www.whitehouse.gov/wp-content/uploads/2022/08/U.S.-Strategy-Toward-Sub-Saharan-Africa-FINAL.pdf.

Wilson, Kalpana. *Race, Racism and Development: Interrogating History, Discourse and Practice*. London: Zed Books, 2012.

Wise, Timothy A. *Eating Tomorrow: Agribusiness, Family Farmers, and the Battle for the Future of Food*. New York: The New Press, 2019.

Wolfe, Patrick. "Settler Colonialism and the Elimination of the Native." *Journal of Genocide Research* 8, no. 4 (December 2006): 387–409. https://doi.org/10.1080/14623520601056240.

Wood, Ellen Meiksins. "The Agrarian Origins of Capitalism." In *Hungry for Profit*, edited by Frederick H. Buttel, Fred Magdoff, and John Bellamy Foster, 23–41. New York: Monthly Review Press, 2000.

World Food Prize. "Dr. Borlaug's CV." https://www.worldfoodprize.org/index.cfm?nodeID=87450&audienceID=1.

———. "WFP Founder Norman Borlaug Receives America's Highest Civilian Honor." https://www.worldfoodprize.org/index.cfm/87428/40024/wfp_founder_norman_borlaug_receives_americas_highest_civilian_honor.

Younis, Musab. "To Own Whiteness." *London Review of Books*, February 10, 2022. https://www.lrb.co.uk/the-paper/v44/n03/musab-younis/to-own-whiteness.

Index

Adesina, Akinwumi, 39
Africa, 1–5; Alliance for Food Sovereignty in, 130; as "battlefield of tomorrow, today," 122–25; as geography of perennial crisis, 109, 126; and matter of empire, 135–37; philanthrocapitalist gaze and, 51–53; private seed sector, 47–51; smallholder farmers in, 89; taking drought gene to, 92–96; Trump administration "Africa strategy," 124–25; US military presence in, 123–24; yield gap and, 45–47; as zone of experimentation, 117–22
Africa, Green Revolution in, 3, centrality of maize in, 83; Gates Foundation's influence in, 59; historicizing, 61; as increasingly financialized, 108; leaders evoking Borlaug, 38; and multinational seed companies, 49; and need for technology, 37; questioning underlying narratives of, 10; roots of, 63; scope of, 8; shaped by philanthrocapitalism, 45; smallholder farmers central to, 41; US power and, 5. *See also* Bill and Melinda Gates Foundation
African Agricultural Technology Foundation (AATF), 8–9, 87, 89, 95–99, 101–2, 104, 131–33, 160n63
African Development Bank, 39

African smallholder farmer. *See* smallholder farmer
"African-led," projects as, 131–33
Agent Orange, 10
agricultural financial technology (agrifintech), 108–9; Africa as "battlefield of tomorrow, today," 122–25; financial experimentation and, 117–22; racial geography and, 117–22; thinking beyond, 125–26
Agricultural Research Organization, 102
Agricultural Survey Commission, Rockefeller Foundation, 65–71
Agriculture and Climate Risk Enterprise (ACRE), 108, 114, 116, 118–19, 125
"Agriculture for Impact," report, 118–19
agroecology, 128–31
Ahtone, Tristan, 72
Alliance for a Green Revolution in Africa (AGRA), 41–43, 46, 48, 50–51, 54, 131–32
Alliance for Food Sovereignty, 130
American Corn Belt, 52, 81, 88
American exceptionalism, 12, 135
American Experience, 24–25
American Hybrid Corn Industry Research Conference, 76
American Seed Trade Association, 47–48
Anderson, Edgar, 77

185

186　INDEX

Appel, Hannah, 45
archival research, 7–8
Arlington, VA, 93
Arvin, Maile, 77
Aschoff, Nicole, 59

Bacillus subtilis, 86
Baldwin, James, 34
basis risk, 115
"battlefield of tomorrow, today," Africa as, 122–25
Baumann, Werner, 130
Bayer, 50, 88, 102–4, 104–6, 108, 129–30, 132
Bhandar, Brenna, 91
Biden administration, 137
Bill and Melinda Gates Foundation. *See* Gates Foundation
"Bill Gates Should Stop Telling Africans What Kind of Agriculture Africans Need," 130–31
biotech crops, 3, 8, 14, 37–39, 41, 87, 138; drought gene and, 92, 95; final frontier for, 104–6; regulatory systems and, 100–2; TELA and, 102–4; testing and commercialization of, 99–100. *See also* genetically modified (GM) crops
biotech seeds, 99, 135, 152n1
biotechnology, agricultural, 2, 8, 10, 13–14, 36, 62, 83, 87–88, 96, 101, 104, 132, 135
Bloomberg Businessweek, 49
Board of Trustees, Rockefeller Foundation, 71
Bolton, John, 124
Borlaug Fellowships, USDA, 18
Borlaug, Julie, 38
Borlaug, Norman, 1–5; adopting Malthus, 29–31; agricultural technology promotion, 36–37; Borlaug Dialogue, 10–11; as brand hero, 7, 16–19; as "father of the Green Revolution," 7, 26, 31, 137–38; first television appearance of, 25–28; lecture of, 31–35; life as prototypical American success story, 20–21, 24–25; portraying as innovator, 138–39; recollecting Great Depression, 27–28; reframing family history of, 21–23; reliance on biblical tropes, 28–29; remembering, 19–25; returning to, 137–39; settler memory and, 23–24; settler-into-immigrant narrative of, 21; studying, 5–12; in US-Dakota War, 21–23; white gaze of, 31–35; Whiteness of, 31–35
Borlaug, Ole and Solveig, 21
Bradfield, Richard, 65

Britain, enclosure of common lands in, 30; financialization of slavery in, 121
Bruyneel, Kevin, 23, 145n14

Canada, farmer groups in, 102
Camacho, Manuel Ávila, 65
capacity building: acquiring Monsanto, 102–4; property relations, 98–100; regulatory systems, 100–102
capitalism, centrality of race to, 59; and frontiers, 119; Gates as storyteller for, 53; Green Revolution and, 24; history of, 61; history of uneven development, 58; improvement and private property, 90; market failures, 43; millennial, 120. *See also* philanthrocapitalism, racial capitalism
Cárdenas, Lazaro, 70
Cargill, 11, 48, 80
Carson, Rachel, 36
Carter, Jimmy, 37
Central American Maize Improvement Program, 79
CGIAR, 80, 117
Chakravartty, Paula, 120
Chesterfield, MO, 128-129
climate change adaptation, 53, 85, 126
climate crisis: anticipating climate shocks, 107–9; building resilience, 109–12; developing security state, 122–25; financial experimentation, 117–122; racial geographies, 117–22; risk management, 112–17; thinking beyond agri-fintech, 125–26
climate risk, 14, 108, 114, 119, 121–23
climate shocks: anticipating, 107–9; risk management for, 112–17
Cold War, 3, 4, 13, 89, 108, 112, 137
Cole, Teju, 135–37
collective amnesia, 35
Cooper, Melinda, 110
Cornell University, 65
corporate consolidation, 49–51
Corteva Agriscience, 50, 81, 108, 132, 135
CRISPR-Cas9, 84–85, 135
cspB, 86
Cutler, Hugh Carson, 77

Dakota people, 22, 38
DDT, 6, 36
decontextualized seeds, 128–31
Dekalb, 79–80, 154n23
Department of Agriculture, US, 18, 65, 68, 132

INDEX 187

Department of Homeland Security, US, 111–12
derivatives, 116–17. *See also* index insurance
Des Moines, Iowa, 1–7, 16, 41, 81, 107, 136
development impact bonds, 121
Discovery Center, Gates Foundation, 54–60
Dow, 50, 162n3
Dr. Norman E. Borlaug World Food Prize Hall of Laureates, 16
drought gene, 86–88; Bayer-Monsanto deal, 102–4; capacity building for, 96–104; licensing, 160–61n63; property relations and, 98–100; regulatory systems and, 100–102; taking to Africa, 92–96
Drought Tolerant Crop Initiative, 93
DroughtGard, 88, 105
Dunbar-Ortiz, Roxanne, 20
DuPont Pioneer, 11, 44, 49–50, 81, 84, 93–94, 129, 149n4
Dyer, Richard, 34–35

ecological debt, 126
economic systems, 110–11
Edge, Mark, 105
Ehrlich, Paul, 30, 33, 138
ejidos, 70
empire, as analytic, 4; bringing into agriculture and climate conversation, 15; Green Revolution and US, 24; intersections with race and finance, 109; in relation to history, 5; and soft power, 111; and US pivot to Africa, 126; and US policy in Africa, 135–37
"Ending World Hunger: The Promise of Biotechnology and the Threat of Antiscience Zealotry," 36
Essay on the Principle of Population (Malthus), 30–31
Essex, Jamey, 112
Ethiopia, 103
eugenics, 68, 78

Famine 1975! (Paddocks), 30
farmers. *See* peasant farmers; smallholder farmers
Ferguson, Rod, 73
Ferroni, Marco, 107
financial crisis, global (2008–9), 117, 120
financial experimentation, racial geography of, 117–22
financialization, 116–17, 119, 121
First the Seed (Kloppenburg), 47
Food and Agricultural Organization, 36

Foreign Policy, 137
Foucault, Michel, 142n6
Fry, Suzanne, 109–10, 122–23

Gates Foundation, 60–61, 62, 87, 159n45; answering call of Borlaug, 41–53; Discovery Center of, 54–60; pedagogies of poverty of, 54–60; philanthrocapitalism of, 43–45; philanthrocapitalist gaze, 51–53; seed companies and, 47–51; taking drought gene to Africa, 92–96; yield gap, 45–47
Gates, Bill, 1–2, 15, 43–46, 136–38; inspiration from Borlaug, 38–41
Gates, Melinda, 1, 45, 54, 56–57
genetically modified (GM) crops, 2–3, 10, 36, 39, 83, 86, 88, 92, 94–95, 99–102, 106. *See also* biotech crops,
genomic editing technologies, 83–85, 135
Global Food Security Strategy, 112–13, 119, 121–22
Global Food Security Symposium, 110, 122–23
global material inequalities, thinking about, 127–28; "African-led" projects, 131–33; agroecology, 128–31; decontextualized seeds, 128–31; matter of empire, 135–37; profit motive, 133–35; returning to Borlaug, 137–39
Global North, 4, 122, 126, 128
Global South, 1, 3–4, 13, 33–34, 36, 58, 108, 120, 126, 128
Global Trends, 109
global war on terror, 123–25, 137
global white ignorance. *See* philanthrocapitalist gaze
Glover, Kaiama, 57, 118–19
glyphosate, 103
Goldstein, Jesse, 134–35
Great Depression, recollections of, 27–28
Green Revolution, 1–5; archival research on, 7–8; backing, 38–61; climate crisis and, 107–26; improvement story and, 89–90; interview-based research on, 8–10; origin story of, 62–85; outlining studies on, 12–15; participant observation, 10–11; remembering, 16–37; seed characterization, 90; studying, 5–12; textual analysis of written and visual materials, 11–12; thinking about global material inequalities, 127–39; yield gap and, 45–47
"Green Revolution, Peace, and Humanity, The" (Borlaug), 31–35

188   INDEX

Grewal, Inderpal, 111
growth chambers, 128
Guardian, 19
Gutterson, Neil, 84–85

Haiti, 118, 136
Hall, Stuart, 30
Haraway, Donna, 154n30
Harrar, George, 79
Harvard University, 65–66
Hawai'i, 129
Hayek, Friedrich, 110
Heart of Darkness (Conrad), 118
Herdt, Robert, 159
Hi-bred Corn Company, 65
Hildyard, Nicholas, 30
Holling, C.S. 110
Horsch, Rob, 96
Howard, Philip, 50
How Europe Underdeveloped Africa (Rodney), 35
How to Avoid a Climate Disaster: The Solutions We Have and the Breakthroughs We Need (Gates), 46, 62, 138
hybrid seed industry, 80, 154n23
hybrid vigor, 47, 65
hybrids, 47–48, 64, 79–80, 82, 84

impatient optimism, 149n7. *See also* Gates, Bill
improvement, ideology of, 88–92
index insurance, 107–8, 113–17, 120–22
India, 1, 25, 33, 74, 91, 104
Indian wars, 21
indigenous peoples, 20–23, 29, 63–64, 68, 72–78, 82, 91
intellectual property, 8, 14, 64, 84, 86, 97–99, 132
International Maize and Wheat Improvement Center (CIMMYT), 62–64, 74, 79–85, 87, 89, 93–94, 97–98, 106
International Rice Research Institute, 90
interview-based research, 8–10
Iowa, 1, 18–26, 41, 141n2
Iowa State University, 74

Jamaica, 80
Johnson, Elmer, 79
Johnson, Leigh, 115–17

Kansas, USA, 86, 88
Kalibata, Agnes, 41–43
Kent Lawrence, 47–48

Kenya, 8, 62, 74, 83–84, 87, 100–103, 107–8, 129, 132–33
Kerry, John, 39, 149n4
Kim, Jim Yong, 39, 149n4
Kish, Zenia, 121–22
Kloppenburg, Jack, 47, 80

land-grant universities (LGUs), 71–74
land, debates over, 71–74
Latin America. *See* Mexico, Green Revolution in; potential (in Mexico), surveying
Lee, Robert, 72
Leroy, Justin, 121–22
Levy, Jonathan, 114
Lincoln, Abraham, 22
Locke, John, 90
Louisiana Purchase, 23

M-Pesa, 108
Maize Lethal Necrosis (MLN), 83
maize, genetically modifying, 62–64
maize, hidden wealth of: collecting samples, 74–75; exchanges of knowledge, 79–82; "firsting" logic, 76–77; possessive logic toward maize, 77–79; scientific detachment, 75–76
Malthus, Thomas, 29–31
Mangelsdorf, Paul, 65
Mann, Charles C., 21–23, 27
Martin, Abby, 127
Martin, Randy, 125
Marx, Karl, 40
McCann, James, 52
McGoey, Linsey, 44
McKinsey and Company, 44
McKittrick, Katherine, 119
Melamed, Jodi, 119–20
memory project, Green Revolution, 16–19; agricultural technology promotion, 36–37; portraying Borlaug as spokesperson, 25–31; remembering Borlaug, 19–25; white gaze, 31–35
mestizos, 69–70, 78
Mexican Agriculture Program (MAP), 7, 63, 70–74, 77–80
Mexico, Green Revolution in: focus on maize, 74–82; genomic editing technologies, 83–85; surveying potential, 65–74; training scientists, 71–74; transforming maize, 62–64
Mills, Charles W., 35, 52
"Minnesota Roots of the Green Revolution," exhibit, 6

INDEX   189

miracle wheat, 25, 146n19
modernization, 29, 34, 47, 70, 82
Monsanto, 3, 8–11, 44–45, 48–49, 62; Bayer acquiring, 102–4; capacity building for drought gene, 96–104; decontextualized seeds, 128–31; improvement ideology, 88–92; property relations, 98–100; public-private partnerships and, 86–88; regulatory systems, 100–102; seed companies, 86–88; taking drought gene to Africa, 92–96; WEMA expansion, 104–6
Montenegro de Wit, Maywa, 130
Moore, Jason W., 119
Moreton-Robinson, Aileen, 77
Morrill Act, 72–73
Mozambique, 87, 96, 100, 103, 152n1
Mukherjee, Roopali, 58
MunichRe, 108, 114

Nader, Laura, 12
Nairobi, Kenya, 7–9, 41, 50, 62, 87, 95, 103, 108
narrative, Borlaug hero, 138; agricultural technology promotion, 36–37; Green Revolution's power as, 4; hybrid seed conversion, 82; portraying Borlaug as spokesperson, 25–31; power of "African led," 132; prevalence of, 16–19; remembering Borlaug, 19–25; of "vanishing Indian," 76; white gaze, 31–35
National Agricultural Research Centers, 87
National Agricultural Research Systems, 83–84, 96, 98, 102
National Biotechnology Development Agency (Nigeria), 104
National Geographic, 58, 118
National Intelligence Council (NIC), 109–10, 122–25, 163n6
natural systems, 110–11
Nazi Germany, 68
neoliberalism, 110–13
Neo-Malthusian, 147n36
Neo-Malthusianism, 29–31
New York Times, 19–20, 43, 145n19, 146n21
Nigeria, 103–4
Nobel Lecture, 31–35
Nobel Peace Prize, 2, 16, 31, 36
Norman E. Borlaug International Symposium, 10
Northrop King, 79, 154n23

Obama administration, 60, 109, 123
Obama, Barack, 6

O'Brien, Jean, 76
Olsson, Tore, 25

Pakistan, 1, 25, 33
Pannar, 49
Paradox of Progress, The, 109–11
participant observation, 10–11
Patel, Raj, 119
patents, 84, 86–87, 97, 129, 135, 156n57, 161n65
PCBs, 10
peasant farmers, 4, 28–30
Pelosi, Nancy, 127–28
permanent adaptability, 111
philanthrocapital, logic of, 45, 51, 134
philanthrocapitalism, 43–45, 59, 88
philanthrocapitalist gaze, 51–53
Philippines, the, 90
Pioneer Hi-bred Corn Company, 65
Pioneer. *See* DuPont Pioneer
Planetary Improvement (Goldstein), 134–35
Plant Physiology, 36
Poor Others, 58
Population Bomb, The (Ehrlich), 30, 33, 138
Population Monster. *See* Borlaug, Norman
postrace, 58–59
potential (in Mexico), surveying: developing "friendly neighbor" policy, 65; focus on farm animals, 67–68; heredity of people, 68–69; land problem, 65–66; new hybrid race, 69–71; possibility of cultural development, 69–70; reporting on "contrasts," 66–68; training scientists, 71–74
poverty capital, 120
poverty knowledge, 40, 53
poverty, pedagogies of, 54–60
Powell, Tracey 103
private seed sector, expanding, 47–51
profit motive, 43–44, 133–35
Program for Africa's Seed System (PASS), 48, 50
project archive, 7
property regime, 87–88, 92, 96, 99, 101, 106, 132, 135
Prophet of Wheat. *See* Borlaug, Norman
public-private partnerships, 8, 11, 13, 87–88, 92–95, 103, 106
Pula, 108, 114, 116, 119, 125

Quinn, Kenneth, 148n3

race, as central to capitalism, 59, 119; failure to "see" in climate crisis, 61; as

race (*continued*)
  geographical, 117; and history of finance, 122; "hybrid," 69–70; ideas of biological fixity, 68; and language of animality, 34; language of biological inheritance, 72; and logic of subprime borrowers, 120; and naturalization of material differences, 58; as produced through property, 91; taxonomies of plants and humans, 78
Races of Maize in Mexico: Their Origin, Characteristics, and Distribution (Wellhausen), 77–78
racial capitalism, 13, 61, 118–20, 136, 139
racial geography, 117–22
Raikes, Jeff, 56
regulatory systems, 87, 100–102, 105
resilience, building, 109–12
resilience, risk management and, 112–17
risk management, 112–17
risk, term, 114
Rockefeller Foundation, 45, 63–65, 71–74, 78–79, 89–90, 93, 98, 131, 143n16
Rodney, Walter, 35
Roosevelt, Franklin D., 65
Roundup Ready, 99, 105
Roy, Ananya, 40, 120
Russia, 124, 137

Saldaña-Portillo, María Josefina, 117–18
Sanders, Bernie, 40
Sasakawa Africa Association (SAA), 37
Sasakawa, Ryoichi, 37
Schickler, Paul, 49
Schnurr, Matthew, 46
Schurman, Rachel, 98
scientists, training, 71–74
security state, developing, 122–25
Seattle, WA, 8, 9, 13, 44, 53, 54
seed banks, 83–85
seed companies, private sector, 47–51
"Seed Security for Food Security," symposium, 81
seeds, 37, 41, 43, 87, 90, 92; banks, 83–85; biotech seeds, 99, 135, 152n1; decontextualizing, 128–31; drought gene, 92–96; governing as private property, 99–100; hidden wealth of maize, 74–82; hybrid seeds, 13, 47, 61, 79, 85, 94, 108, 114–15, 158n17; landraces, 13, 62–64, 82–85; maize seeds, 57, 80–81, 87; Monsanto and, 102–6; philanthrocapitalist gaze and, 51–53; possessive orientation toward, 64; private sector of, 47–51; securitizing

smallholder farmers, 107–8, 114–15; yield gap and, 45–47
settler colonialism, 12, 20, 22–24; and land grant universities, 73; and logic of possession, 77; race and property integral to, 91; violence of, 139
settler memory, 21–24, 29
Shah, Raj, 60
Shulman, George, 146n20
Sierra Club, 36
Silent Spring, 36
Silva, Ferreira da, 120
slavery, and extraction of Africa's "resources," 53; financialization of, 121; violence of, 139
smallholder farmers, 2–4, 8, 11–14, 41, 43, 48, anticipating climate shocks, 107–9; building resilience, 109–12; developing security state, 122–25; financial experimentation, 117–22; risk management for, 112–17; thinking beyond agri-fintech, 125–26
social impact bonds, 121–22
South Africa, 49, 87, 92, 94, 96, 100, 103–4, 129, 152n1
Stakman, E. C., 65, 71–72, 143n16
Standard Oil, 70
State Department, US, 11, 112
St. Louis, 8, 95–96, 128–29
Strategic Futures Group, NIC, 109–11
Structural Adjustment Policies, 48
Sturken, Marita, 19
sub-Saharan Africa, 2, 47, 87, 89, 92, 95, 100, 104, 132, 137
subprime, logic of, 120–21
SwissRe, 108, 114
Syngenta, 48, 107–8, 129, 132
Syngenta Foundation for Sustainable Agriculture, 83, 107–8, 116

Tanzania, 87, 95–96, 100, 103, 152n1
techno-optimism, 53–54, 126
TELA. *See* Water Efficient Maize for Africa (WEMA)
Thailand, 74
Trump administration, 124–25
Turse, Nick, 123–24

"U.S. Strategy toward Sub-Saharan Africa," document, 137
Uganda, 87, 96, 100, 102–3
Ukraine, 137
United Nations Climate Change Conference, 40

INDEX   191

United Nations Convention on Climate Change International Meeting, 127
United States, 4–5, 7, 20, 80, 99, 102, 109; agricultural biotechnology in, 87–89; and "battlefield of tomorrow, today," 122–25; comparing to Mexico, 63–65, 70; countries of strategic importance to, 109–12; debates over land use, 73; drought gene in, 92–93; farmers in, 46, 53; "friendly neighbor" policy with, 63–65; matter of empire, 135–38; as "nation of immigrants," 20–23; outside borders of, 59–60; racial geography and, 117–22; risk management, 112–17; as settler colonial country, 20–21; trade agreement with Kenya, 132–33; white gaze and, 31–32, 36; and White Savior Industrial Complex, 135–37; yield gap, 46
University of Minnesota, 5, 7, 9, 27, 65
US Agency for International Development (USAID), 3–5, 62, 83, 87, 93, 111–12, 117, 132–34
US Chamber of Commerce, 132
US Department of Agriculture (USDA), 18, 68, 132
US Global Food Security Act, 112–15
US Global Food Security Strategy, 115–16
US Intelligence Community, 109, 124
US Office of Trade Representative, 133
"U.S. Strategy toward Sub-Saharan Africa," 137
US-Dakota War, 21–22

Vietnam, U.S. war in, 32
Villegas, Daniel Cosío, 66
Vine, David, 21
vulnerability, building resilience in world of, 109–12

Walker, Jeremy, 110
"Walking with Water," exhibit, 58
Wall Street Journal, 103, 144n3
Wallace, Henry A., 65
Washington Post, 136
Water Efficient Maize for Africa (WEMA), 8–9, 11, 63, 83; adopting TELA as new name, 152n1; capacity building for drought gene, 96–104; continuing under Bayer, 102–4; expansion of, 104–6; property relations, 98–100; public-private partnerships and, 86–88; regulatory systems, 100–102; taking drought gene to Africa, 92–96; yield improvement, 88–92
Wellhausen, E. J., 74–80
white gaze, 31–35
white ignorance, 35, 138
White Savior Industrial Complex, 136
Whiteness, 12, 31–35, 59, 70; and possessive logic, 77
Wilson, Kalpana, 34
Wise, Tim, 27
World Bank, 117
World Economic Forum, 40, 130
World Food Prize Conference, 1–2, 41, 81, 107, 141n2; participant observation at, 10–11. *See also* Norman E. Borlaug International Symposium
World Health Assembly, 40
World War II, 34, 65
Wynter, Sylvia, 119

Xolocotzi, Efraín Hernández, 75–76, 154n27

Yaqui Valley, Mexico, 26
yield gap, 45–47, 52, 100
yields, improving, 88–92

Founded in 1893,
UNIVERSITY OF CALIFORNIA PRESS
publishes bold, progressive books and journals
on topics in the arts, humanities, social sciences,
and natural sciences—with a focus on social
justice issues—that inspire thought and action
among readers worldwide.

The UC PRESS FOUNDATION
raises funds to uphold the press's vital role
as an independent, nonprofit publisher, and
receives philanthropic support from a wide
range of individuals and institutions—and from
committed readers like you. To learn more, visit
ucpress.edu/supportus.